THE BINDING OF THE NILE

AND

THE NEW SOUDAN

Mehemet Ali,

from a painting in the possession of the Oriental Club.

LONDON: EDWARD ARNOLD 1904.

THE BINDING OF THE NILE AND THE NEW SOUDAN

BY THE HON.

SIDNEY PEEL

LATE FELLOW OF TRINITY COLLEGE, OXFORD

NEGRO UNIVERSITIES PRESS
NEW YORK

Originally published in 1904
by Edward Arnold, London

Reprinted 1969 by
Negro Universities Press
A Division of Greenwood Publishing Corp.
New York

SBN 8371-2369-0

PRINTED IN UNITED STATES OF AMERICA

PREFACE

I HAVE tried to tell in outline the story of the regulation of the Nile and some of its consequences. A rash project, perhaps, for one who is not an engineer; but, then, this book is not written for engineers, and politics enter largely into it.

I have had some special opportunities of observation, and I have many friends to thank for the help which they have given me. In particular I am much indebted to the *Standard*, whose special correspondent in Egypt and the Soudan I had the good fortune to be during a part of 1902-1903.

Anyone who wishes to gain a real acquaintance with the principles and details of Egyptian Irrigation should read the monumental and interesting work by Sir W. Willcocks, K.C.M.G., on that subject, to which my indebtedness is large.

The standard work on the Soudan is not yet written. There is but one man who combines the necessary knowledge, experience, and attainments to do it, and he, fortunately for the Soudan, is—and will be for a long time to come—too fully occupied with his arduous and multifarious duties. I mean, of course, the present Governor-General, Sir Reginald Wingate. No official of the Soudan Government, least of all he, has leisure to write a book; but I sincerely hope that the materials for it are steadily collecting.

S. P.

CONTENTS

PART I

THE BINDING OF THE NILE

PART II

THE NEW SOUDAN

CONTENTS

PORTRAIT

PART I

THE BINDING OF THE NILE

CHAPTER I

THE NILE

Far back in the world's history a fracture of the earth's crust took place in the region which is now Egypt, and the sea filled the valley as far as a point not much north of Assouan. Into this fiord ran several rivers from the high ground east and west, bearing down with them heaps of detritus, and forming small deltas like the plain of Kom-Ombos. On the sea-bottom were laid down deposits of sand and gravel, and then the land began to rise. Meantime the volcanic movements of East Central Africa had shaped the country into its present configuration, and the rivers which drained from the great lakes and swamps of the south, and those which flowed down from the high plateau on the east, combining their waters somewhere about Khartoum, pushed their marvellous course northwards, and began the creation of the fertile soil of Egypt. From this time onwards the climatic conditions must have continued very much what they are to-day. Changes, of course, there have been in the level of the land; but the sea-valley had become a river-valley, and year by year the annual flood increased the cultivable soil, and refreshed it with moisture, just as it would be doing to-day if left untrammelled by the devices of man.

Late in the history of the river-valley, but very early in the history of humanity, this favoured strip of country became the home of men, who doubtless cast their seed upon the slime left by the retreating waters, and reaped their crops long before the dawn of history. It is remarkable and characteristic of the conditions of the country that tradition ascribes to the earliest King of Egypt, Menes, the first King of the First Dynasty, the first attempts to regulate the flow of the river—in other words, the first scheme of irrigation proper.

If it were possible to divert the river from its course, and effectually to bar its way before it reached the boundaries of Egypt, what an appalling catastrophe would follow—no mere disaster, but absolute annihilation! On the coast lands of the Mediterranean a sparse population might still eke out a miserable existence by storing the scanty rainfall, but nowhere else. The very oases of the desert would be dried up, and in a short time the shifting sands of the Sahara would have overlaid the deposits in the river-valley, and buried out of sight even the ruins of the past. The waters of the Nile are, and ever have been, the sole giver of all life in Egypt.

Whoever finds himself in Cairo should lose no time in taking his stand upon the bridge, and in reflecting upon the history of the water that goes sliding and eddying beneath him, on the way to perform its last duties among the cotton-fields of the delta. Some of it has been travelling for three months from its sources beyond the Victoria

Nyanza, itself over 1,100 metres* above sea-level. From the Victoria Nyanza it has passed down the Somerset Nile into the Albert Nyanza, thence a five days' journey to Lado, past Duffile and the Fola Rapids. From Lado to Bor the fall is still rapid, but henceforward as far as Khartoum, some 1,000 miles, the stream is on a very feeble slope. Between Bor and the junction of the Ghazal River on the left bank is the region of the sudd, floating masses of compressed vegetation, which, if neglected entirely, block the course of the river. Here, too, are the wide and desolate marshes, inhabited by myriads of mosquitoes and that strange, melancholy bird, the whale-headed stork (*Balæniceps rex*).

The Ghazal River contributes very little to the flow of the Nile, owing to wide lagoons through which it passes, and which cause great evaporation. The river then passes sharply to the right, and sixty miles further on is joined by the Sobat from the eastern hills, which in flood brings down a volume equal to that of tho Nilc, but during summer contributes little or nothing. From the white sediment brought down by the Sobat the White Nile derives its name and colour. Hence it flows in a wide bed, a mile across on the average, 540 miles to its junction with the Blue Nile at Khartoum. Khartoum is still 1,800 miles from the sea, and 390 metres above it. Two hundred miles north of Khartoum, near to Berber, the Atbara River flows in, and hence the Nile pursues its solitary way through the desert until, after a circuitous bend round Dongola, it bursts through

* 1 metre = about 39 inches.

the rocky defiles of Nubia, and emerges at Assouan into Egypt proper. Of the distance between Khartoum and Assouan about 350 miles consist of so-called cataracts, during which the total drop is 200 metres; about 750 miles are ordinary channel, with a total drop of nearly 100 metres.

Of all the affluents of the Nile, the Blue Nile (assisted by its tributaries, the Rahad and the Dinder) and the Atbara have been of infinitely the greatest importance to Egypt in the past. Not only do their waters contribute the largest proportion of the annual flood, but also down them comes the rich volcanic detritus swept from the Abyssinian hills by the heavy summer rains, which composes the red-brown silt so dear to the Egyptian cultivator. To understand the system of irrigation in Egypt it is necessary to have a clear view of the amount of water derived from the different sources at different times of the year.

The great lakes and swamps of Uganda, acting as reservoirs, prevent any great differences in the discharge of water above Lado. At that place the low Nile discharge is about 500 cubic metres* per second; but, in spite of the Ghazal River and the Sobat, so great is the loss by diffusion in the marshes and by direct evaporation, that at Khartoum the discharge is no more than 300 cubic metres per second. At this time the Blue Nile is giving no more than 160 cubic metres per second at Khartoum, and the Atbara is not running at all. The loss between Khartoum and Assouan is about 50 cubic metres, and consequently the amount of water passing Assouan in May in an ordinary year

* A cubic metre of water equals, roughly, 1 ton.

is 410 cubic metres per second. This is the summer supply of Egypt.

Meantime heavy rains have been falling about Lado and on the Sobat. Towards April 15 the river begins to rise, the effect of which is felt at Khartoum about May 20, and at Assouan about June 10. A most curious phenomenon accompanies this preliminary increase—the appearance of the 'green' water. It used to be thought that this 'green' water proceeded from the sudd region. During the previous months the swamps in that country have been lying isolated and stagnant under the burning tropical sun; their waters have become polluted with decaying vegetable matter. When once more the rising river overtops its banks, this fœtid water was supposed to be swept out into the stream, and finally make its appearance in Egypt. But a closer examination of the facts, which has only been possible since the Soudan was re-opened, has caused this view to be abandoned.

The 'green water' is caused by the presence of an innumerable number of microscopic algæ, which give it a very offensive taste and smell. So far as can be ascertained, their origin is in the tributaries of the Sobat above Nasser. The rains in April carry them out into the White Nile, and thence they pass down to Nubia and Egypt. Under a hot sun and in clear water they increase with amazing rapidity, and sometimes they form a column 250 to 500 miles long. These weeds go on growing, drying, and decaying, until the arrival of the turbid flood-water, which at once puts an end to the whole process, as they cannot survive except in clear water.

Horrible as the 'green' water is, its appearance is hailed with delight by the Egyptian, for he knows that it is the forerunner of the rushing waters of the real flood-time, and the first sign of the coming close of the water-famine. By September 1 the Nile at Lado has reached its maximum of 1,600 cubic metres. In the valley of the Sobat the rains last till November, with the result that by September 15 or 20 the White Nile at Khartoum has attained its maximum discharge of 4,500 cubic metres per second.

Meanwhile great things have been happening on the Blue Nile and the Atbara. About July 5 the Blue Nile begins to rise, and the flood comes down with considerable rapidity till it reaches its maximum of 5,500 cubic metres per second on August 25. The famous 'red' water reaches Assouan on July 15, and is seen ten days later at Cairo. The flood on the Atbara would begin at nearly the same time as on the Blue Nile, but for the fact that it spends a month in saturating its own dry bed and the adjoining country. Once it begins, however, about July 5, it comes very rapidly, and about August 20 reaches its maximum, which is usually some 3,400 cubic metres per second, but occasionally amounts to as much as 4,900.

If these three contributors, the White Nile, the Blue Nile, and the Atbara, all reached their maximum at the same time, the result in an ordinary year would be a discharge at Assouan of some 13,000 to 14,000 cubic metres per second, and in occasional years a very great deal more. But this is not the case. In an ordinary year the Nile is at its lowest at Assouan at the end of May, discharging no more than 410 cubic metres per second. After the arrival of the 'green' water it rises slowly till July 20. By that time the 'red' water is fairly on the move, and the rise goes on with increased rapidity till the maximum is reached, on September 5, of 10,000 cubic metres. But both date and amount are liable to variation. If the White Nile flood is a weak one, and the Atbara early, the maximum may be reached a day or two earlier; but if the White Nile is very strong it will not be attained until September 20. A late maximum, in other words, means a good supply of water in the White Nile, and as the White Nile is the principal source of supply after the flood is over, this fact is of inestimable importance to Egypt. But occasionally, as in 1878, this is carried to excess. In that year the White Nile flood was very late and very high. The maximum was not reached till September 30, when 13,200 cubic metres per second were passing Assouan, with very disastrous results in Egypt. All through the following summer the supply was very good, and at its lowest, in May, was more than three times as great as the average.*

* For the purpose of illustration, it is interesting to compare the discharge of the Thames at Teddington:

	Cubic Metres per Second.
During June the average discharge for the twenty years ending 1902 was	35
The average in June, 1903, was	178
The discharge on June 21, 1903, was ...	387
On February 21, 1900, it was	533
And on November 18, 1894 (greatest on record), it was	1,065

I have given the discharge in cubic metres per second, the unit generally in use on the Nile. On the Thames the

Curiously enough, the preceding year offered a startling contrast. It was the lowest Nile on record. The maximum was reached on August 20, and was some 3,500 cubic metres below the average, and in the following May the discharge at Assouan fell to 230 cubic metres per second, as against the average 410.

By the end of October the Atbara has usually disappeared altogether, and the Blue Nile falls very quickly after the middle of September. The White Nile, however, owing to the regulating effect of its natural reservoirs and the slackness of its current, is very much more deliberate in its fall, and the results at Assouan are as follows in a normal year: By the end of September the discharge has fallen to 8,000 cubic metres per second; end of October to 5,000; end of November to 3,000; end of December to 2,000; end of January to 1,500; end of March to 650; and end of May to 410.

In other countries the year is divided into seasons, reckoned according to the position of the sun and the temperature, or according to the rainfall. In Egypt the state of the all-important river is the principal factor in determining the seasons. These are, first, the months of the inundation, or Nili, August to November; second,

---

figures are usually given in gallons per day, which sounds much more imposing. If the number of cubic metres per second is multiplied by about 1,900,000, it gives approximately the number of gallons per day. But, after all, the discharge of the Thames in June, 1903, was not so very far below that of the Nile during the same month.

the winter, or Shitwi, December to March; and, third, the summer, or Sefi, April to July, when the river is at its lowest. Nowhere else do the actual seasons correspond with the nominal in a manner so regular and unfailing. But, of course, within these limits there is a great variation in the amounts of both the maximum and the minimum of the volume of the Nile, and the dates at which they occur from year to year. During the twenty-six years 1873-1898 the maximum flood at Assouan varied from the exceptional height of 9·15 metres above zero on the gauge to the equally exceptional 6·40 metres. And the dates on which these maxima were attained varied from October 1 in the first case to August 20 in the other. The dates on which the volume of water passing through reached its minimum varied from May 8 to June 24; while the worst lowest on record was ·71 metre below zero, and the best was 1·88 metres above it. (Zero, it should be explained, is the lowest level which the river would touch in an average year.)

These figures are the result of the free and unimpeded flow of the Nile. To us, with our wider field of observation, with our knowledge of the sources of the Nile, and preciser information as to the conditions prevailing in those distant countries, the behaviour of the river seems, after all, but the resultant of many natural causes, and capable of prediction, and even regulation. But to those who, living in a country where rain was practically unknown, knew nothing of the tropical lakes or the rain-shrouded hills of Abyssinia, and who merely saw the great river issuing from the burning deserts of the south in flood at the very time when other rivers were parched and dried, how marvellous it must have seemed, and how inexplicable must have appeared those occasional variations, threatening destruction, on the one hand by excessive inundation, and on the other by famine and drought! Few floods in the history of Egypt can have been higher than that of 1878, and few lower than that of 1877, when 1,000,000 acres were left without water. But several great famines have been recorded in the past. Perhaps the worst of these began in A.D. 1064. A succession of low Niles took place, lasting for seven years, like the seven lean kine of Pharaoh's dream. Terrible results followed. Even human flesh was eaten, and the Caliph's family had to flee to Syria. Similar disasters occurred in A.D. 1199.

Such extraordinary catastrophes are rare, but their possibility is always present to the minds of those responsible for the government and welfare of Egypt. We shall see what steps have been taken to guard against them, and first we shall examine the different systems of irrigation which have prevailed, and the nature of the services which the water, whose variations we have described, is made to perform before it finally reaches the sea.

CHAPTER II

BASIN IRRIGATION

THERE are two kinds of irrigation in Egypt—basin and perennial irrigation. Basin irrigation is the ancient and historical method of the country. Tradition ascribes its invention to the first King of Egypt, and it is obviously designed to take full advantage of the annual flood. Practically the same as it was 7,000 years ago, it may be seen in Upper Egypt to-day.

The cultivable land from Assouan to the Delta is, with the exception of the Fayoum, that altogether remarkable province, a narrow strip of country, varying in breadth from a few miles to nothing at all, sometimes on both sides of the river, sometimes only on one. Being of deltaic formation, the land is highest near the river, and slopes away towards the desert. As the river flows from south to north, there is, of course, also a general slope of the country in the same direction. Earthen dykes are run at right angles to the river as far as the desert, a dyke parallel to the river and close to its bank connects them, and so a basin is formed, enclosed on the fourth side by the desert. Thus the land is arranged in a series of terraces. Usually these basins are arranged in a series, one basin draining into its neighbour, the last of the series discharging back again into the Nile. Sometimes a second dyke, parallel to the river, divides the lower land near the desert from the higher; sometimes the arrangement is still further complicated by other dykes, making enclosures within the area of the original basin.

The object of the basins is to regulate the supply of the flood-water. Each series of basins has special feeder canals to lead into them. These are shallow, and have their bed about halfway between the high and low level of the Nile. They are therefore dry in the winter and summer, and only run during the flood. The heads of these canals, where they take off from the Nile, remain closed by dams or by masonry regulators till the silt-bearing flood is coming down in sufficient strength. Then about August 10 or 12 the canals are opened and the basins filled. The lowest basins in each series are filled first, then the next lowest, and so on. In a low flood, as in 1877, there is not enough muddy water to go round, and the upper basins get no water at all; such lands are called 'sharaki,' and are exempt from taxation. For forty days the flood-water stands on the land, thoroughly soaking it and washing it, and at the same time depositing its fertilizing silt. At the end of that time, through the escape at the lower end of the series, the water is discharged back again into the Nile. But if the flood is a very high one or very slow in abating, the date at which the water can be discharged and the basins dried has to be postponed. Fortunately, this seldom happens; but when it does it has a doubly bad effect. The oversoaking is said to engender worms, and also the ripening of the crops is postponed to an unfavourable season of the year.

Against this particular evil there is no remedy, but since the British occupation a great deal has been done to improve the system of basin irrigation, and prevent a large amount of sharaki even in years of low Nile. These measures consist in arrangements for distributing the 'red' water more evenly over the whole system, and not allowing the lower basins only to receive the full benefit, while the higher merely receive 'white' water—*i.e.*, water which has already deposited its silt. Further, since any water is better than none, the systems of basins have been so managed, where possible, that the water from an upper system can be drained into a lower one, and thus make up deficiencies, though with water of an inferior quality. More attention has also been paid to the angle at which the feeder canals take off from the Nile, and the slope on which they are laid, so as to provide as much 'red' water as possible, while diminishing the amount of silting up. The result of this work is seen in the diminution of the sharaki lands year by year. In 1877 these amounted to nearly 1,000,000 acres, a loss of £1,000,000 in land-tax. During the next ten years the average was 45,000 acres, an annual loss of some £40,000. In 1888 the loss was £300,000, and the Egyptian Government expended £800,000 in remedial works, which have had an extraordinary effect. In 1899, when the Nile was exceptionally low, the sharaki lands amounted to 264,000 acres only, as against nearly 1,000,000 in 1877.

As for the high lands lying immediately between the river and the basin dyke, only eight or nine times in a century does the Nile rise high enough to flood them. They are called 'berms,' and are ingeniously irrigated by means of special high-level canals, which, starting from a point above the head of the basin system, or perhaps leading down from an upper system, pass by means of a siphon under the feeder canal. The berms are, of course, also irrigated by lifting water directly from the Nile itself.

There are forty-five systems of basins in Upper Egypt, most of which, and those the largest, are on the left bank. Some of the feeder canals are insignificant, and feed only two or three basins. Others, like the Sohagia Canal, south of Assiout, feed an extensive system, and are real rivers when full. The basins themselves are 5,000 to 15,000 acres in extent, and it can easily be understood that when they are taking in they have an enormous effect in diminishing the pressure of the flood, and, on the other hand, their discharge lengthens it out in the lower reaches of the river, when the level has already fallen very much at Assouan. While the basins are filling in August and September, they absorb about 2,000 cubic metres per second. Besides this, a considerable amount is employed in filling the channel of the Nile itself and its branches. Evaporation, absorption, and direct irrigation, also play their part, and the result is that the discharge of the Nile at Cairo is some 2,500 cubic metres less than it is at Assouan. But during October and November the basins are discharging. The southernmost ones are empty by October 15, those in the neighbourhood of Cairo not till about November 30, or even later. The consequence is that the Nile at Cairo in October is discharging 900 cubic metres per second more than at Assouan, and in November 500 more.

In November, therefore, the visitor to Cairo can still get some idea of what Upper Egypt is like in flood-time. From desert to desert it stretches, one vast lake, divided by a network of dykes, and studded here and there with villages raised on artificial mounds, which year by year rise higher on their own ruins. A greater flood than usual makes terrible havoc in these villages; for the rising water soon crumbles their mud walls, and the whole collapses like a pack of cards. Every mortal thing is living on the dykes, which play the part of roads; only the water-fowl, emerging in thousands from their secluded marshes, spread themselves in security over the wide waters, and here and there an isolated villager may be seen in an ancient palm-wood tub, paddling and baling by turns. The dykes run on the same lines as they have run for centuries, and on them the traveller, on his way to visit the Pyramids of Gizeh or the Tombs of the Kings at Sakkarah, the burial-ground of ancient Memphis, may watch the whole life of Egypt pass and repass in long procession, set as in a frieze. Work in the fields, of course, is at a standstill, but the villages are humming with preparation for the sowing, and alive with flocks of goats and sheep, camels, buffaloes, and asses; even the rats have been forced to the same refuge, and may be seen popping in and out among the roots of the palm-trees.* Where one basin drains into another the fisherman spreads his net, and reaps a rich harvest in the rush of the current. At intervals are stationed pickets of grave watchers, squatting patiently alongside large bundles of millet or maize stalks. This is all that remains of the corvée service, and gladly is it borne.

For every villager is interested in the preservation of the dykes, and apart from ordinary accidents this great lake, owing to the swift changes of temperature in the neighbouring desert, is liable to violent storms, which drive great waves against the crumbling dykes, and would soon break them if left alone. The millet stalks are put down to break the force of the waves. Each village omdeh, or headman, is responsible for these arrangements, and he, too, may be met upon the dykes, a picturesque figure in flowing black and white, mounted on an ambling Arab pony, going round to see that all his sentinels are on duty. How near the past is brought, when you enter the tombs and find painted on the walls, or figured out in stone, the same people engaged in the same pursuits, as though Egypt had not been since then for thousands of centuries coveted and seized in turn by so many invading nations! The cultivator of the soil, moulded by the unchanging and imperious demands of the great river, to which he owes his whole subsistence, has retained the customs, and even the

* The Egyptian peasant, however, refuses to accept the prosaic evidence of his eyes about these rats, and, like the stout conservative he is, prefers to believe the old tradition that they turn to mud during the flood season. Many a man will gravely assert that he has himself observed the transformation actually in progress.

features, of those remote forerunners, who are his ancestors in everything, except, perhaps, by descent of blood.

Under the Pharaohs and under the Romans the whole of Lower as well as Upper Egypt was under basin irrigation, and the whole country was cultivated. In those days Egypt was the granary of the Mediterranean, and at the time of the Arab conquest, A.D. 700, her population was estimated to number 12,000,000. Under Arab rule began that period of deterioration which lasted for 1,100 years, and which, had the system of irrigation been less natural to the physical conditions of the country and less simple, would have resulted in an absolute abandonment of cultivation, and reduced Egypt to the state in which Mesopotamia, once the garden of the world, finds itself to-day. Even as it was, by A.D. 1800 the population was brought down to about 2,000,000, and all the northernmost and greater half of the Delta had become a neglected and uncultivated swamp. War, famine, and pestilence, in turn, had played their part; but the fundamental cause of all was the misgovernment, which had neglected the irrigation. For in basin irrigation, as, indeed, with all irrigation, two things are of the utmost importance: the first, to get the water on to the land; and the second, to drain it off again. Salt is the great enemy to be fought. Not only do the Nile waters contain a large quantity of salts, in solution, but the strata underlying the alluvial deposits, being of marine origin, are also rich in salts. If the water is allowed to stand on the land, evaporation takes place, until nothing but a salt efflorescence is left.

While if the land be so water-logged that the level at which water can be obtained by digging is brought near to the surface, the water containing salts from below is drawn upwards by means of capillary attraction, and once more evaporation takes place, leaving the salts in the soil. It is clear that as the natural level of the land approaches sea-level it becomes more and more troublesome to provide proper means of carrying off the water. Accordingly, the northernmost parts of the Delta were the first to suffer, and gradually the line of cultivation receded.

Nor was this all. No one looking at a map of the Delta can fail to have been struck by that extraordinary feature of the northern coast-line, the great lakes. At the present time there are in the Delta about 3,430,000 acres of land cultivated or under reclamation, and another 500,000 acres of waste land. North of these lie 1,180,000 acres, either permanently covered by the lakes or else flooded by them from September to December. Between the lakes and the sea is a belt of sand-dunes or sandy plains, pierced occasionally by openings. The sand-dunes are constantly being augmented by the prevailing north-west winds. These lakes or lagoons are for the most part extremely salt, and are distributed as follows, beginning from the west: Lake Mareotis, 70,000 acres; Edku, 60,000 acres; Borillos, 180,000 acres, and as much more during flood and early winter; and Lake Menzalah on the east, largest of all, 490,000 acres, and flooding 200,000 acres more at the same time of year. All these waste lands, now known as the Berea, were cultivated in Roman times, some being occupied by vineyards, others by wheat, and it would seem that the lakes were kept from extending landwards by dykes. But when the land was allowed to go out of cultivation no one had any interest in looking after the regulation of the lakes. First, the passage through the sand-dunes became silted up, because, as the basins decreased in number, less water was drained from them into the lakes at the time of the inundation. Then, after the closing of the openings, the water gradually rose, breached the neglected dykes, and completed the ruin of the land. Once the openings in the dunes are closed, the lake has to rise to a considerable height before it can force its way through again, owing to the continuous action of the sand driven by the wind. In this way it came to pass that neither of the lakes had more than one opening into the sea, and consequently, rising in flood-time above sea-level, invaded the lands to the south.

Another cause may have possibly contributed to the same effect—namely, a sinking of the coast lands. Nowhere is this a more probable explanation of the facts than in the neighbourhood of Lake Manzalah. Before the Arab conquest much of what is now a shallow lake was famous for its gardens, palm-groves, vineyards, and wheat-fields; besides its agricultural villages, it contained towns famous for their cloth and cutlery manufactures, like Tunah, Damirah, Dabik, and, above all, Tinnis. But at the time of the conquest these towns were already islands in the lake, a position which enabled them to be the last stronghold of Coptic resistance to the Moslems.

Whatever the cause or the combination of causes—and the history of these tracts remains very obscure—the results were the retreat southwards of the cultivated area, with the consequence that, after over 1,000 years of Mohammedan rule, Egypt found herself in the weakened and impoverished condition already described; saved only from annihilation by the system of basin irrigation, of which the traditions survived, though in a diminished area, stubbornly preserved from age to age by her industrious and conservative peasantry.

CHAPTER III

PERENNIAL IRRIGATION

It was the Viceroy Mehemet Ali who revolutionized the methods of Egyptian agriculture, and introduced what is known as perennial irrigation —that is to say, irrigation all the year round, as opposed to irrigation during the flood only. In all the annals of the East there are few more striking figures and few histories more exciting than that of the Albanian tobacco-seller, who, rising high in the favour of Sultan Mahmoud, was sent to Egypt as Viceroy in 1810. Adopting a method well known in Cairo, and well calculated to secure the respect, and even the affection, of Orientals, he consolidated his power by the treacherous murder of the Mamelukes, and thoroughly organized the military resources of his province. Summoned to the aid of the Sultan, his armies bloodily stamped out the successful rising of the Greeks in the Peloponnese, and left to himself he would have settled the question of Greek independence once for all. But his summary proposal to transport the whole Christian population, and repeople the Morea with Mohammedan plantations, provoked the intervention of the Powers, and at Navarino he suffered the complete loss of his fleet. Undismayed, he conceived the idea of making himself master of the Turkish Empire. His armies overran Syria, and easily overcoming the Turkish opposition, he threatened Constantinople itself. Once more, if left to himself, he would have succeeded in his object, but once more the slow processes of European diplomacy at last resulted in action. British guns gave the death-blow to his hopes at Acre, and Turkey was saved from her ambitious vassal, though Egypt remained a practically independent State, and her sovereignty became the hereditary appanage of the house of Mehemet.

In the region of domestic policy this strange combination of barbarism and genius proved that he retained the commercial instincts of the tobacco-seller as well as the far-reaching ideas and the drastic methods of the despot. He perceived the advantages which would accrue from the cultivation of cotton and the sugar-cane, hitherto unknown in Egypt. These crops are impossible under a system of basin irrigation; for though they require to be watered all through the summer, they would be ruined by complete inundation, and the shallow flood canals are well above the summer level of the river. But difficulties were nothing to Mehemet Ali. The corvée was called out, and the unfortunate fellaheen were set to work to dig new canals and reconstruct the existing water-ways in the Delta, so as to render them capable of conveying water during the period of low Nile. At the same time the dykes along the banks of the Nile and the canals were very much strengthened, so as to keep out the flood; the old basin dykes were obliterated, and arrangements made for irrigating the land from the new canals.

Of course, perennial irrigation in itself had always existed in Egypt. It would have been indeed strange if the principle applied by anybody who daily waters a window-box had not occurred to the Egyptians. The Nile berms were often enclosed to protect them from inundation, and watered directly from the river all the year round, while within the basins themselves considerable tracts were irrigated from wells. But never before had special canals been provided by Government for perennial irrigation. The advantages of perennial irrigation are that crops like cotton and sugar can be grown, which would otherwise be impossible, and that two, or even three crops, can be produced in a year, instead of only one. The land is therefore increased in value, but, on the other hand, there are serious objections. First of all, the land is deprived of the full benefit derived from the annual renewal of the soil by the silt deposit. Agriculture becomes a much more intricate and difficult process; the exhausted soil has to be constantly refreshed and stimulated by dressings and manures. The basin irrigator makes less profits, but he has less risk and less anxiety; he can only ride a donkey, while the perennial irrigator can ride a horse. —But behind the horseman sits black Care. A low summer supply means to him the waste of many weeks' labour and much expenditure in preparing and sowing his fields; the basin irrigator does nothing till the flood is over, and should the inundation not cover a part of his land, it merely means that that part lies fallow for another year, and suffers no deterioration. A breach in the dykes during the flood is inconvenient to the one, no doubt, but fatal to the other, for it means the ruin of his growing cotton or sugar. And under perennial irrigation it is much more likely to occur, for the basins act as a safety-valve in the inundation, and, while they lengthen out the period of the flood in Lower Egypt, enormously decrease its volume at any given time; but when the same lands are receiving only an occasional watering, the volume that rushes to the sea is by so much the greater, and the pressure on the dykes is heavily increased.

If Mehemet Ali had been content to preserve the old basin dykes, the vivifying effects of the flood-water might have been occasionally applied, and some of these dangers averted. But, as we have seen, everything had to give way to the immediate cultivation of cotton, and the dykes were levelled. Nor was this the only error committed. The new canals were faulty in slope and alignment. Too often their subsidiaries were constructed merely with the object of carrying water to the lands of powerful and favoured individuals, without regard to the general interest. It was found in consequence that an enormous silting up of the canals took place every year. But the Viceroy, with all the forced labour of Egypt at his free disposal, took little heed of this, and vast numbers of men were dragged from their homes every year to redig the canals. Even so it was impossible for the task to be completed before the next flood came round. The lower reaches of the canals remained choked with mud and weeds, and, worst of all, proper drainage was neglected.

When Napoleon was in Egypt in 1798, his master mind, accustomed to go to the root of many matters in spite of all the alarms and distractions of war, perceived how much might be done by a regulation of the water at the point of the Delta. His idea was to close each branch of the river alternately during the flood, and so double its effect. Mehemet Ali proposed to apply the same principle to low-water, and began to close the head of the Rosetta or left-hand branch of the Nile with an enormous stone dam, so as to divert the whole supply into the Damietta branch. Linant Pasha, then chief of the French engineers, who had been brought over to advise upon the new works, persuaded him to abandon this design, and proposed instead to build a barrage upon each branch, constructing them in the dry, and then diverting the Nile into its new course by means of earthen dams. The Viceroy approved, and at once proposed, with characteristic energy, to dismantle the Pyramids and make use of their material, just as Sultan Hasan had once stripped the marble casing from the Great Pyramid to construct his famous mosque. Fortunately, the prosaic question of transport arrested this design, and new limestone quarries were opened near Cairo instead. But although workshops were built, material collected, and foundations dug, Mehemet seems to have lost all interest in the work, and in 1835 he abandoned it altogether, and for seven years nothing more was heard of it.

Two things appear to have operated in his mind. It seemed simpler to keep on digging out the main canals by the help of the corvée, and cheaper, too, because nothing was wanted but the unpaid labour, though it was false economy. And, secondly, his ill-regulated but far-reaching imagination was already busy upon a new idea, nothing less than the construction of a great reservoir, which should store up the surplus water of the winter and let it out again in the summer, thereby, as he supposed, rendering unnecessary the construction of any minor work like a barrage.

Surely there can be no more curious instance of the irony of fate than the history of these two great ideas—the Barrage and the Reservoir. Both in turn have been carried into successful execution by engineers and statesmen belonging to the very nation which shattered the ambitions of their first authors. Against the crumbling walls of Acre, held so stoutly for weeks by the English sea-captain, when even the notion of a day's defence seemed a madness, Napoleon's dreams of Eastern empire dashed themselves vainly to pieces. Forty years later the same walls could not withstand for half a day the guns of their former defenders, and Mehemet Ali in turn saw his own imperial dreams finally dissolve. Mehemet neglected the Barrage, because he thought the Reservoir would make it unnecessary; and yet in the end it was only the successful working of the Barrage which gave new life to the project of the Reservoir, and made its completion an absolute necessity.

The Viceroy, perhaps, deserves little credit for his idea. In matters of irrigation it is often much easier to see what to do than how to do it; like other great men, he imagined better than he knew. Undoubtedly he was fascinated by the fame of

King Amenemhat of the Twelfth Dynasty, and his construction of Lake Mœris in the Fayoum. He constructed a new regulator at the spot where the Bahr Yusuf enters that extraordinary province, and even built himself a house there. But whether because of the inherent difficulties of the subject, or because of foreign complications, nothing was done until in 1842 his mind reverted to the idea of the Barrage at the point of the Delta. The Frenchman Mougel had the address to couple his design with a scheme of military fortifications, and, attracted by this double advantage, Mehemet at once ordered the works to be begun, though, unfortunately, his energy was more devoted to making the point of the Delta the military capital of Egypt than towards the regulation of the water. By this time, however, Mehemet Ali's career was drawing to a close, and in 1848 he died, leaving the work in which he never more than half believed still unfinished.

It is difficult to conceive anything more humiliating and exasperating than the position of the French engineers who acted as advisers to the Egyptian Government. Time has vindicated their reputation, and proved the excellence of their designs and the soundness of their work; but in their own day they had to suffer disappointment, and even disgrace, and to bear the brunt of failure, due not to themselves, but to the conditions under which they lived. With no authority to enforce the execution of their plans, hampered at every turn, sometimes by the incompetence, and always by the unwillingness, of the Arab engineers through whom they had to work, supported only occasionally by the uncertain breeze of viceregal caprice, they struggled bravely on, and deserve the greatest credit for what they did manage to accomplish. In 1853 Abbas Pasha, the then Viceroy, dismissed Mougel from his service, to mark his displeasure at the slowness of the building, and appointed a new man. Little was gained by the change. The Barrages were nominally finished in 1863, and an attempt was made to close the gates on the Rosetta branch. But a settlement took place, and they had to be immediately reopened. Not till 1872 was the Barrage really used, and then only partially on the Rosetta branch, and not at all on the Damietta. Still, whereas before 1872 only 250,000 acres of summer crops had been matured in the Delta, and that at the cost of enormous labour in clearing the canals, afterwards the total was 600,000 acres, and the cost of maintenance was very much less. The ordinary summer supply available for the Delta canals was increased from 64 cubic metres per second to 150.

This success brought home to the mind of the then reigning Khedive, Ismail, the advantages of perennial irrigation and the cultivation of cotton, and he determined to extend the system to Upper Egypt and the Fayoum, where he possessed huge estates, amassed by fair means and foul through the agency of the notorious Mufettish, Said. Accord ingly, in 1873 the great Ibrahimiyah Canal was dug. Starting from near Assiout, it runs for 268 kilometres nearly parallel to the Nile on its left bank, and supplies perennial irrigation to 252,000 acres in the provinces of Assiout, Minia, and Beni-Suêf.

It also carries flood-water to a series of basins lying to the west of it, nearer the desert. Before 1873 the Bahr Yusuf, which feeds the Fayoum, took its water direct from the Nile, but its head was now transferred to the left bank of the Ibrahimiyah Canal at Dêrut, and 327,000 acres in the Fayoum came nominally under perennial irrigation.

It would perhaps have been more reasonable to perfect the irrigation system of Lower Egypt, and to complete the Barrages entirely, before embarking on new projects in Upper Egypt. But the temptation to improve his own lands by simply calling out the corvée to dig canals was too strong for Ismail; and, indeed, he was not the man to devote himself to the carrying out of old projects to the exclusion of new ones. In him the vigorous and practical originality of his grandfather Mehemet appeared in the form of a fantastic imagination running riot in all directions, unrestrained by the prosaic considerations of time and means. Yet with able Ministers he might have been one of the greatest of rulers. In spite of all the degradation which his reckless extravagance brought on Egypt, the country owes him something; for there was generally something great in his ideas, and time is carrying many of them into effect. It is impossible not to feel some admiration for the man who, when asked what gauge the Soudan railway should be, replied, 'Make it the same as that of the railways in South Africa. It will save trouble in the end.'

CHAPTER IV

THE CULTURE OF THE FIELDS

It has already been explained that on the Nile berms or high banks, which are covered by the flood only once in six or seven years, on islands in the river, and on selected tracts within the basins in the neighbourhood of wells, it has been the immemorial custom to lift water on to the fields. Everywhere the two primitive instruments of ancient Egypt are in common use to-day—the shadoof and the sakieh. The shadoof is a long pole balanced on a support. From one end of it is suspended a bucket, and from the other a heavy counterpoise, equal in weight to the bucket when full of water. The bucket is made of various materials, very often leather, though the ordinary kerosene-oil tin of commerce is making its presence felt here as elsewhere. The shadoof is worked by hand. The bucket is pulled down into the water, then lifted up by the help of the counterpoise, and its contents are tipped over into the channel leading to the cultivated land, where the water is steered by means of miniature canals and dams into the required direction. I suppose there could not be a simpler form of unskilled labour than working the shadoof. Whenever I think of the fellah of Upper Egypt, I think of the shadoof. Up and down, creak and splash, hour after hour, day after day, he goes on lifting and tilting, with an amazing and monotonous regularity. Nothing disturbs him—not even a steamer grounded on a sand-bank twenty yards in front of him. As he stands, naked except for a loin-cloth of blue cotton, with his absolutely dull, impassive features, his magnificent chest and arms but weak legs, you cannot help wondering which came first, the shadoof or the shadoof-man, so perfectly are they adapted to each other. Two piastres are the humble guerdon of the long day's labour. You can calculate upon the fellah as you could on a machine. But, in spite of it all, deep down in his soul lies the sentiment which redeems him and distinguishes him from a mere machine—his absorbing love for the soil. Take him away, set him to other tasks—to serve, for instance, in the army—he will perform his duties with the same unfaltering regularity and docility; but all the time he is thinking in his heart of the black soil and the water of Egypt. In Omdurman I asked a more than usually intelligent Egyptian soldier, who had been told off to perform some small services for me, 'Do you like being in the army?' Without hesitation came the answer, 'No.' 'What do you want to do?' 'I wish to be at home,' he said, 'and cultivate the ground.'

With a single shadoof water can be lifted 2½ metres; but when the bank is high a second or third tier of shadoofs is employed, and in some places as many as five shadoofs may be seen lifting the water from one level to another, till it reaches the fields. One man working twelve hours a day can lift enough water to irrigate an acre of cotton or corn in ten days.

The sakieh, or Persian water-wheel, consists of a vertical wheel with a string of buckets attached to it, which, as the wheel turns round, are let down into the water, come up full, and discharge their contents into a channel as they come to the top. The wheel is turned by means of spokes, which catch in a horizontal wheel worked by oxen, buffaloes, or some other beast of burden. If the lift is high, the string of buckets may be very long; but if the wheel itself dips into the water, there may be no string at all, and it is then called a taboot. The buckets are often earthenware pitchers, and the wheels themselves are generally of the rudest construction, and made of palm wood; but new and improved iron water-wheels are coming into use. Still, whether of iron or of wood, they all seem to make the same peculiar sing-song whine. There is no sound more characteristic of Egypt. It has a peculiar penetration. Night and day it continues. I believe Egyptian music is founded upon it. The fellaheen say the cattle will not work unless they hear it. Certainly when one stops, the other stops also.

In the Fayoum, where, owing to the difference in the levels, the canals have often a very high velocity, there are very ingenious water-wheels or turbines, which play the part of sakiehs, but are turned by the force of the current; the water thus lifts itself continuously.

Where the lift is very little, shadoofs and sakiehs are replaced by instruments called Natalis and Archimedean screws; but, naturally, since the introduction of perennial irrigation has so increased the area to be watered by lift, machinery has had to be called in, and most of the work is done by pumps worked by steam. Each large land-owner has his own pump and engine, which can be moved from place to place, and are also hired out to the smaller men. On very large estates stationary engines have been erected, which, of course, are able to raise a much larger amount of water; but as a rule the machinery employed is a portable eight-horse-power engine and an eight-inch centrifugal pump.

How different is all this from the lot of the agriculturist in other lands! For him there is no digging or maintenance of canals; no apparatus of regulators, dams, sluice-gates, siphons, and drains; no painful lifting of the water by pumps and engines, shadoofs, and sakiehs. The rain falls upon his fields from heaven without any effort of his. He looks to Providence to regulate his supply; the Egyptian looks to a Government department. But the Egyptian, as a compensation for his extra labour, has the advantage of greater certainty. He knows the sun will shine. The rise and fall of the Nile, variable as it is, can be foretold with greater exactness than that of any other river—with far greater exactness than the duration of the rainy season in any country in the world. Nature indeed made his task simple in the extreme, if he had been content with one crop a year. Every year the flood thoroughly washed the land, and kept it free from injurious salts; it also covered them with a deposit of mud, which relieved him from the necessity of dressing and manuring the exhausted soil. The Nile silt, though singularly rich in potash, the principal food of leguminous plants, like peas, beans, and clover, is, however, very poor in the nitrates on which cereals depend. But the Egyptian clover, called bersine, has the property of secreting nitrates from the air, and depositing them in the soil to an extraordinary extent, so that the land was able to bear crops of clover and cereals in rotation to an unlimited extent without any manuring. The desire to grow rich by crops like cotton and sugar, and by forcing the land to double its output, has changed all this. Not only has the summer supply of water become of the utmost importance, but the soil has to be constantly refreshed with manures. The question of manures is, indeed, only second, under perennial irrigation, to the question of water.

Wherever cattle and stock are numerous, farm-yard manure is used, as well as the guano from the immense colonies of pigeons, kept for the purpose in specially built pigeon-lofts throughout Egypt. Between Halfa and Kena there are in-exhaustive supplies of nitrates in the desert, and north of Kena the mounds which mark the sites of ancient cities, like Abydos, Ashmunên, Medinet, and the rest, serve the same purpose. The ruins of the past are thus valued by the agriculturist not less than by the archæologist, perhaps even more so; for lands in proximity to them are rented higher in consequence. Year by year more attention is now paid to the dressing of the soil, as perennial irrigation is more understood and more studied; besides the natural resources of the country, an increasing amount of manures is imported from abroad, and there is little doubt that the growing tendency in this direction will continue.

But all preparation of the soil is worse than useless labour unless the necessary amount of water can be provided. This amount varies both with the nature of the crop, the season of the year, and the position of the land. The critical time is, of course, the summer, when the supply of water is least and the heat is greatest, and, of course, in Upper Egypt, where the sun is strongest, and the loss by evaporation consequently greater, the demand is more urgent than in the Delta.

The total amount of cultivable land in Egypt is 6,250,000 acres. Before the completion of the new works, to which period all the figures in this chapter refer, the total nominally under cultivation was about 5,750,000 acres. Of this, Upper Egypt claimed 2,320,000—viz., 587,000 nominally under perennial irrigation, and 1,732,000, including 1,435,000 under basin irrigation every year, and 297,000 of Nile berms, which are only flooded once in six or seven years, and at other times irrigated directly from the Nile by means of shadoofs and water-wheels. Of the whole of this area only about 20 to 30 per cent. produce double crops in the year; for the amount of perennial irrigation is but small, and, although the whole of the Fayoum —329,000 acres—was supposed to be perennially irrigated, so faulty was the water-supply that the summer crops were only about 30 per cent. of the whole, instead of 50 per cent., which is the rule in the Delta. It seems, indeed, as though matters had been arranged expressly for the benefit of the tourist; it is Upper Egypt in the winter season that he goes to see, and it is then that the fields are green with corn, clover, and other crops. The following table shows the different crops, and the acreage devoted to them at the different seasons in Upper Egypt:

Season.	*Acreage.*	*Crops.*
Summer ...	372,500	Sugar, cotton, vegetables, melons, summer sorghum (or millet).
Flood ...	530,000	Flood sorghum, rice.
Winter ...	2,120,000	Wheat, beans, clover, barley, lentils, flax (little), onions, vetches.

In Lower Egypt, or the Delta, the total area of fully or partly cultivated land is 3,430,000 acres, and there are still 500,000 acres of totally unreclaimed land. All this is under perennial irrigation; half of it is under summer crops every year, and 40 per cent. produces two crops a year. The following is a similar table to that given for Upper Egypt:

LOWER EGYPT.

Season.	*Acreage.*	*Crops.*
Summer ...	1,674,000	Cotton, sugar, vegetables, rice.
Flood ...	980,000	Maize (nearly all), rice.
Winter ...	2,139,000	Wheat, barley, clover, beans, vegetables, flax.

Even if we look only to the summer acreage under crops, it is obvious that the water-supply in the summer is very important; but when we look to the value of the crop it becomes much more striking. Far the most valuable crop in Egypt is the cotton, which is the principal item in the summer. In Upper Egypt the value of a summer crop is, on the average, more than twice that of a winter crop per acre, and in the Delta the proportion is nearly the same. And though the value of the flood crops is increased by the date-palms, of which there are 5,700,000 paying taxes in Egypt, and whose produce is gathered at this season, it cannot be compared with the summer crops. Sugar-cane is now but little grown in the Delta, and even in Upper Egypt its acreage is rather less than that of cotton.

Valuable as are the cotton-plant and the sugar-cane, it must never be forgotten that one of the humbler winter crops, though valued much lower in point of money, is yet the foundation of well-being in the others—I mean the Egyptian clover, or bersine, the friend of beast and man alike. Long before I knew its remarkable properties, I admired it for its beauty. Green and glossy, it covers acre upon acre with a luxuriant carpet, in pleasing contrast both to the black soil and the desert sand, and most refreshing and comforting to the eye. A bundle of it will satisfy even the grumbling camel ; even the melancholy buffalo looks a shade less depressed when her turn comes to be tethered in it for her meal. Sheep and donkeys can hardly eat it down fast enough within the circle of their ropes before it has grown up again. And all the time it is steadily collecting in the soil the invaluable globules of nitrate, which will put new life into the succeeding cotton or corn. The part it plays in preparing the soil can be estimated by the rotations of crops followed by the Egyptian cultivator. These are as follows:

ON RICH SOILS.

	Winter.	*Summer or Flood.*
First year	Clover	Cotton.
Second year	Beans or wheat ...	Indian corn.

ON POOR SOILS.

	Winter.	*Summer or Flood.*
First year	Clover	Cotton.
Second year	Clover	Cotton.
Third year	Barley	Rice or fallow.

Rice and barley have their place, because they are less affected by the injurious salts, which are the great enemies of the soil's fertility.

In Lower Egypt cotton is sown from the end of February to the beginning of April. The land is well watered before it is ploughed for the seed, and again when the seed is sown. From then until the beginning of the flood it is watered on the average about once in twenty days. The harvest lasts from August 20 to November 10, and the cotton is picked two or three times over. During this time the crop is watered about once in every fifteen days, but as the water is now abundant there is nothing to fear. Indian corn is sown from July 5 to August 30, and October 15 to November 30 is the period of harvest. It is irrigated at the time of sowing, twenty days after, and then once in ten or twelve days. The first two of these waterings are, of course, the important ones. The earlier it is sown, the better the crop will be, because it will have better weather for maturing; but if the flood is late, and consequently the water-supply is low, the Government may have to resort to a system of rotations in sending water down the canals, and then the Indian corn crop may be sacrificed to the interests of the cotton. Rice is the wettest of all the crops; the kind (called 'sultâni') sown in May and reaped in November is watered once in ten days before the flood, but during the flood is given as much water as the drains can carry off. The other kind (called 'sabaini') is sown in August, and also reaped in November. Both in Lower and Upper Egypt it is purely a flood crop, and takes all the water it can get. The winter crops, wheat, beans, barley, and clover, are sown in November and December. Wheat and beans are irrigated twice, barley once, but clover goes on growing up till June, and takes more water according to the number of crops, sometimes three or four, that are taken off it.

In Upper Egypt cotton-sowing begins at the same time, but the harvest is earlier. Sugar-cane is sown in March, and the canes are cut from December 15 to March 15. Sometimes the same roots are left in the ground, and produce another crop in the second year; but this is never of such quality as the first, and the land has probably to be left fallow after it. Sugar, therefore, though nominally more valuable than cotton per acre, is more costly in the long-run. It is watered every twelve or fifteen days. The other crops are the summer and flood sorghum, grown on the berms or in tracts within the basins, and irrigated by shadoofs and water-wheels about once every ten days; and the wheat, beans, clover, and barley, in the basins. The cereals are usually not watered at all, but the clover follows the same course as in Lower Egypt.

Summing up these results, we find that the principal crops in Lower Egypt are cotton and rice. The cotton needs irrigation about once in twenty days, the rice once in ten days. To provide this amount of water, a canal should discharge (after allowance has been made for wastage) 22 cubic metres in twenty-four hours per acre of cotton, and 40 cubic metres per acre of rice. That is to say, 1 cubic metre per second will suffice for 4,000 acres of cotton and 2,150 acres of rice. In Upper Egypt rice is only a flood crop, and cotton and sugar need about 25 to 30 per cent. more water than in the Delta, owing to the greater loss from evaporation—that is to say, 1 cubic metre per second will only suffice for 3,000 acres of cotton or sugar. During the winter the land throughout Egypt requires on the average a watering once in forty days. But, as we have seen, it is the summer supply for the cotton that is the really important thing. We shall see later what the effect of the reservoir is likely to be in safeguarding and extending these interests.

CHAPTER V

THE DELTA BARRAGE AND THE ENGLISH ENGINEERS

At the date of the English occupation the Delta Barrage was generally thought to be like the whole fabric of Egyptian Government, rotten to the core. And so indeed it seemed. No one had ever dared to use, or apparently even to think of using, the Barrage on the Damietta or right-hand branch at all. The history of the Barrage on the Rosetta branch was hardly less inglorious. In 1863 its gates were closed for the first time, but about ten of its arches began to settle, and ominous cracks showed. Eventually the threatened part was surrounded by a coffer-dam, and from 1872 to 1883 it managed to hold up about 1 metre. But even that was precarious. Commission after Commission had condemned the structure; it was felt that at any moment it might give way, especially if called upon to bear a greater strain, and it was actually the settled policy of the Government to rely on huge and costly pumping-stations instead. It was a paltry result after the expenditure of £4,000,000 and so much labour.

Then, not for the first or last time, the Anglo-Indians came to the rescue of Africa. Sir Evelyn

Baring himself (now Lord Cromer) during his service as Financial Member of the Council in India, must have been impressed by the enormous importance of irrigation. It would not be difficult to find many points of resemblance between his character and that of one of the greatest, if not the greatest, of the rulers of India, Lord Lawrence, different as were their spheres of work; but certainly they were alike in this. As Lord Lawrence supported Arthur Cotton in his engineering work, so Lord Cromer supported Colin Scott-Moncrieff and the band of trusty lieutenants—Willcocks, Garstin, Ross, Brown, Foster, Western, and Reid—who came with him. Fortunately for Egypt, these men, trained in the best school of irrigation in the world, possessed not only the highest scientific skill and knowledge, but were also animated by the best spirit of the empire-building Englishman. Deep in them lay the earnest wish and determination, far stronger even than their enthusiasm and love for their profession, to alleviate the lot of the unhappy peasantry of Egypt. It was this heartfelt sympathy for the wrongs of the fellaheen, ground down by the intolerable burden of the corvée, that sustained them in their ceaseless labours and enabled them to pass successfully through those dark days, when the air was full of forebodings of failure and disaster, whose fulfilment would have pleased so many.

The Barrage is situated, as has been said, a little way back from the point of the Delta. It is really two Barrages, one on the left or Rosetta branch of the Nile, with sixty-one arches, 465 metres in length, and the other on the Damietta branch with seventy-one arches, 535 metres in length. Between the two runs a revetment wall across the intervening tongue of land, 1,000 metres in length. From a distance it resembles a bridge of rather fanciful design, with the arches set unusually close together, and, indeed, for a great part of its career the functions of a bridge were the only ones it performed. The tongue of land between has been converted into beautiful gardens, planted with shady trees and many shrubs and flowers, and even a greensward resembling grass. Altogether, it is one of the most delightful and beautiful spots in Egypt, besides being one of the most useful. Here is the starting-point of the great feeder canals which irrigate the Delta provinces. On the left, facing north, is the Rayah Behera, which supplies the province of Behera, to the left of the Rosetta branch. Between the two Barrages is the head of the Rayah Menoufia, the canal which feeds the two provinces of Menoufia and Gharbia, lying between the two arms of the river; while on the right is the Rayah Tewfiki, which, with its supplementary canals, Ismailia, Sharkia, and Basusia, supplies the three eastern provinces of the Delta, Kalyubia, Sharkia, and Dakalia. All these canals are navigable, as well as the branches of the river, and provided with locks for that purpose. These great waterways are free to all, and few of the results of British occupation are more appreciated in Egypt. Formerly all craft upon the Nile had to pay toll on passing under a bridge, which did nothing but hinder their progress, while those for whose convenience it was made passed without charge overhead.

A Barrage, as its name implies, is designed to completely bar the bed of the river, so as to enable it to feed the canals at a higher level than would otherwise be the case, and also to allow the flood to pass through it easily. It needs, therefore, a very solid foundation from bank to bank, on which the arches which hold the movable sluice-gates can be securely planted. Its construction is, therefore, a very different and much more difficult matter than merely throwing a bridge over the stream, even a bridge with several spans. The difficulty is all the greater when, as here, the bed of the river offers nothing more substantial than shifting sands to build upon. It was for this reason that Linant wished to build the Barrages at leisure in the dry, and then divert the river from its old channels, and lead it through when they were completed. But Mougel chose to build his in the existing bed of the river, thereby increasing the difficulties of actual construction, though from other points of view there was much to be said for this plan. At the site of the Rosetta Barrage the bed of the river was not of uniform depth; he therefore filled up the deepest part of the channel, which lay on the right, with loose stones, so as to bring it up to the level of the bottom on the left-hand side. No cement was used in laying down this barrier, but the Nile mud filled the interstices and made it water-tight; when finished, this barrier was 60 metres wide and 10 deep at the deepest part. On this and on the natural sand he built a platform 46 metres in width and 3·5 metres thick, composed of concrete overlaid with brick and stonework. On the platform he raised his arches and piers, all built of brick. Each of the openings for the sluice-gates, sixty-one in number, was 5 metres wide. Like an iceberg, that part of the Barrage which is visible above water is much less than the invisible part below. To further strengthen the structure and keep it in its place, a mass of rubble pitching or loose stones was thrown into the river on the downstream side. This talus was 3 to 16 metres in depth, and at one part extended 50 metres downstream in a kind of tongue, narrowing down to 2 metres. The Damietta Barrage was built on a similar plan, but its downstream talus was not so large. Unfortunately, the concrete used for the platform was inferior, chiefly owing to the fact that Mehemet Ali, growing impatient at the slow progress of the work, ordered a certain amount of material to be laid down every day, and laid down it had to be in defiance of all engineering requirements. The consequence was that, as soon as the Rosetta Barrage was subjected to strain, ten of the arches on the left-hand side, where the platform was laid down on sand only, settled and cracked. It was patched up by surrounding the injured arches with a coffer-dam; but the Damietta Barrage never even had its gates put in.

Such was the structure with which the English engineers had to deal. Even as it stands to-day, it cannot, of course, compare in magnitude with many works upon the Indian rivers; but as regards the difficulties to be overcome, it can compare with almost any in the world. It would have been far easier to rebuild the whole thing from the beginning, but at the time the necessary funds were not forthcoming. They had to take the old structure, with all its imperfections, and screw it up to work as it was. The country could not afford to cut off the summer water-supply of the Delta while the repairs were in progress. The cotton-crop had always to be thought of. And the period of the year during which the summer canals required to be supplied was the only period during which work could be done, for once the flood came down all operations were at an end. It is the glory of the English engineers that, working under these conditions and with untrained workmen, they succeeded in their task.

The Government was already paying many thousands a year to a company for pumping water out of the Rosetta branch into the canals during the summer, and the first thing Sir Colin Scott Moncrieff had to do on his arrival was to decide upon a scheme which had been prepared for erecting new pumping-stations at an initial expense of about three-quarters of a million, and involving an annual expenditure of at least another quarter of a million. So hopeless were the prospects of the Barrage assumed to be, that even this expenditure, with a doubtful result, was thought preferable to repairing it. Sir Colin's arrival was only in the nick of time. He determined to see what could be done with the resources at hand. The new pumping-station scheme was set aside, and Mr. Willcocks was put in charge of the Barrage.

There was much literature on the subject. During the last sixteen years nothing had been done, but much had been written, and more said. Commissions, expert and inexpert, had issued voluminous and condemnatory reports, and had even prepared expensive schemes of repair. Mr. Willcocks (now Sir William) is an indefatigable reader, and could hardly have been encouraged thereby, till an examination of the structure itself showed that all the later reports had been drawn up without reference to facts. It had been observed that whenever the gates were let down there was very severe action of the water on the downstream side. The authors of the reports concluded that the foundations were honeycombed. It is characteristic of the Looking-glass days of Ismail that no one ever thought of trying to find out by actual observation whether there might not be some other cause. But Mr. Willcocks, looking for himself, found that this action of the water was caused, not by honeycombed foundations, but by open gratings which intervened between the bottom of the sluice-gates and the platform. They had been put down originally to keep the silt away from the bottom of the gates. Someone had fixed them so as to prevent the gates from being lowered to their full extent and then they had actually been forgotten. Measures were at once taken to close these gratings, and eventually to remove them altogether. 20,000 cubic metres of rubble pitching were added to the talus. The Damietta Barrage was likewise strengthened with various ingenious expedients, improvised to meet the demands of the moment. Sluice-gates were put in for the first time and gradually closed. Part of it was closed by a temporary stone dam. Eventually in the summer of 1884 2·2 metres of water were held up on the Rosetta branch, and 1 metre on the Damietta. Next year the same nursing process was continued. The coffer-dam round the weak arches was strengthened, the talus of rubble pitching below each Barrage was completed, and this year 3 metres were held up on the Rosetta branch and 1·6 on the Damietta. The effect was extraordinary. The acreage under summer cultivation was doubled, rising from 600,000 to 1,200,000 acres. Not only was the supply of water in the Delta canals greatly increased throughout the summer, but, as it was delivered at a higher level, there was a great saving of expense in lifting it on to the land. For the first time the Egyptians thoroughly realized that a new power had come amongst them.

The experiment had been successful, but temporary expedients could not last for ever. The more water held up, and the greater the area of the summer cultivation, the more necessary it became to insure the stability of the structure. A thorough repair would cost money. Fortunately, this was now forthcoming. Mr. Willcocks' success had settled the claim of the Barrage to a share in the famous Irrigation Million borrowed in 1885.

At the end of 1886 the work was begun, under the charge of Colonel Western and Mr. Reid, sent specially from India for the purpose. The operations were spread over four years. In the first year the left half of the Rosetta Barrage was taken in hand and finished before the flood, next year the right half. In 1889 and 1890 the Damietta Barrage was similarly taken in hand and finished. Each year the part to be repaired was enclosed by earthen dams, and the water pumped out so as to lay the foundations dry. The whole of the existing floor was raised, both on the upstream and downstream side, and it was also considerably lengthened. It was, in fact, enclosed in a new and reliable suit of armour. The dangers and anxieties of the work were incessant. The protecting dams were always liable to be breached. Spring after spring burst out through the treacherous bed of the river, and threatened the destruction of the year's work ; and again and again each of them was successively stopped by a number of ingenious devices. There is no enemy so persistent and so insidious in its attacks as running water. It is always feeling for and finding out the weak spots. It never sleeps or slackens by day or by night. It can only be met successfully by a corresponding activity. While work was possible, it was carried on unceasingly by night as well as by day. Sometimes as many as 1,600 men worked through the night. The upper brickwork was generally sound, but new iron sluicegates moving in special grooves were fitted throughout. The whole of the repairs cost £465,000. It was money well laid out. Not only was the safety of the Barrage assured, but it was found possible to hold up yet another metre of water. The area of summer crops matured rose once more from 1,200,000 to 1,520,000.

It might have been thought that the work was now complete. Both in 1891 and 1892 (a year of specially low summer supply) all the water in the Nile was held up, and diverted into the canals. Not a drop reached the sea during the summer without having done duty. But the engineers were now looking forward to a time when the supply would be greatly increased. The idea of a reservoir had become an affair of practical politics. It was necessary to make assurance doubly sure. Accordingly, in 1896 a new experiment was tried—namely, stock-ramming with clay.

Certain arches in the Damietta Barrage were selected, and in them five-inch holes were bored right down through pier and platform alike. When the bore-holes were complete, they were lined with iron tubes. Clay was then forced through the tube by means of an iron rammer, and as much as could be made to spread out at the bottom of the hole was put in and rammed. As far as the clay went, the experiment was not an entire success; but the boring brought to light a condition of things in the very vitals of the Barrage which demanded drastic treatment, for the bore-holes proved the existence of large cavities in the original platform, and in some places there was free water communication between one bore-hole and another. Some piers in the Rosetta Barrage were therefore chosen for a similar experiment, but this time liquid Portland cement was used instead of clay, and the results were entirely satisfactory. Few discoveries have been of more signal service than the invention of Portland cement. It is not too much to say that it has revolutionized hydraulic engineering by the facilities it affords for constructing solid works in water. Its strength and resisting power is enormous, but its greatest quality is that it hardens and solidifies under the action of water, and, so far as is known, only goes on getting harder and harder with time. The borings in the Rosetta Barrage having revealed similar deficiencies to those in the Damietta, it was decided to apply to both a thorough dose of this invaluable and invigorating medicine.

In 1897 five holes were bored in each pier of the Rosetta Barrage (their united length amounted to very nearly 6 kilometres), and into each was poured a quantity of liquid cement. The necessity for the treatment was proved by the fact that in some cases the cement travelled right through from the bore-hole in one pier and rose through the bore-hole in an adjoining pier till it reached the top. One pier actually swallowed 439 barrels of cement, while its neighbour took a lesser but still gigantic draught of 327 barrels. There was no doubt that the cement thoroughly explored and filled all the cavities existing in the foundations under the bridge. In all, 3,254 barrels were used in the Rosetta Barrage alone. In 1898 the grouting process, as it is called, was applied with equal success to the Damietta Barrage.

To use Sir Hanbury Brown's homely but expressive image, the process applied to the Barrage was exactly that followed by a cook who wishes to finish off a cold pie with its proper complement of jelly. The jelly is introduced into the pie in the form of warm gravy, which penetrates into and fills every recess of the succulent interior, and then solidifies as it cools.

And still the engineers were not satisfied. So treacherous is the river's bed that no possible safeguards seemed superfluous. It speaks volumes for the courage and skill of those who in 1885 held up 3 metres of water with the old unreformed Barrage, that in 1897, after the successful execution of such great and costly repairs, it was still thought advisable to undertake completely new works to assist in the task of holding up 4, or at the most 5, metres.

It is a principle in hydraulics, not easily understood at first by the layman, that the pressure upon a weir or barrage in a river depends entirely upon the difference in level between the water on the upstream and on the downstream side, and not on the mere volume of water in the river behind it. In December, 1897, the Caisse de la Dette voted £530,000 for the construction of two subsidiary downstream weirs, with the object of relieving the pressure on the Barrage by raising the level of the water on the downstream side, thus dividing the head of water to be held up into two—in other words, by making two steps instead of one. Each weir was to consist of a core of rubble masonry set in cement, sunk well below the bed of the river, and protected up and down stream by a long slope of rough stone blocks or pitching. To make the masonry core thoroughly watertight, a mass of clay puddle was to be put on either side of it. The weirs were thus to be a solid dam, blocking the course of the stream up to such a height that the head of water on the Barrage, at that time amounting to 4 metres, would be reduced to 2·5 metres. The flood would pass freely over the top of the weirs. At the same time the sluice-gates of the existing Barrage were to be heightened, so as to permit the upstream level to be raised 1 metre more in June and July, so as to take full advantage of the rising flood and facilitate the early sowing of maize, a great point with the Egyptian cultivator.

By the summer of 1900 this programme had been completed. The building of the weir on the Rosetta branch was an especially fine performance, for which great credit was due to Sir Hanbury Brown and Mr. Brooke, who were in charge of it. Five hundred metres in length, it was begun at the end of December, 1899, and actually finished before the flood began to come down in July. The same Portland cement played a great part in its construction. It may now fairly be said that the Barrage is complete at last, and fully equal to every strain that it can be called upon to bear.

The weirs were constructed not a moment too soon. It so happened that the summer supply of 1900 was lower than in any previous year of which records have been kept. In 1889 the river sank to a level of ·60 metre below zero on the Assouan gauge. In 1878 it fell to ·71 metre below zero, and this was the lowest known before the summer of 1900. But on three days in that year, May 15, 16, and 26, the river fell to a level of ·91 metre below zero. The position was aggravated by the extension of summer cultivation. The total extent of summer crops had risen still further to over 1,700,000 acres.

To save the valuable cotton crop was the earnest preoccupation of the Irrigation Department. They were able by the most strenuous efforts, not merely to save the crop, but so to treat it that it gave a yield which, only a few years before, would have been considered perfectly impossible even in a good year. But all their efforts would have been in vain had it not been possible, thanks to the new weirs, to raise the level of the water upstream of the Barrage to an extent which would have been exceedingly dangerous without their assistance, and so to take full advantage of the rising flood. The mere enumeration of the special measures which were put into force gives a very good idea of the difficult duties which devolve on those who control the water in Egypt :

1. Earthen dams were constructed in both branches of the river to prevent the inrush of salt water from the sea.

2. Special programmes were laid down for 'rotations' on the canals.

The system of rotations, which was introduced from India, is that the land-owners are only allowed to pump water on to their lands at certain intervals. There are several advantages in this. The water is economized, and as it can thereby be kept at a lower level in the canals, there is less danger of the soil becoming deteriorated by excessive saturation. The pumps are allowed to work for a certain period, according to the district, and then an interval is prescribed, until the expiration of which they are not allowed to work again. In 1900 the pumps were allowed to work for a period of six days at a time, and at first twelve days was the interval until the next pumping. But as the summer wore on, and the river continued to fall, the interval was gradually extended to twenty-two days—a very severe measure indeed.

3. All land-owners were warned not to sow rice.

4. They were also forbidden to sow maize until a date should be announced.

5. Special pumping arrangements were made.

6. There was more than usually careful regulation at the heads of the canals above the Barrage, so as to insure a proper distribution of the water available to all the provinces.

7. A special staff was appointed to see that all these regulations were carried out.

Could any government be more paternal than this—it might even be said, more despotic? But countries which depend on irrigation have a natural tendency towards despotism. When water is plentiful they may be as republican and democratic as you please; but when the crisis of scanty water comes they must have a strong hand over them, just as the Roman Republic had to have its Dictator in times of national peril. It speaks well for the good sense of the Egyptians, and it proves their implicit faith, built up by sixteen years' experience, in the English engineers, that even those stringent regulations were unhesitatingly obeyed, and that breaches of them were so rare as to be almost non-existent. They had their reward; for while 1878 is still remembered as a year of black disaster and distress, in 1900 the cotton crop amounted to no less than 5,250,000 kantars,* and the maize crop, in spite of its late sowing, was also very good. Only the rice, a comparatively insignificant item, was sacrificed to its more important rivals. Thanks mainly to the good work done by the completed Barrage, neither the public nor the private finances of Egypt

* 1 Kantar = nearly 100lbs.

suffered the least shock from a year of unprecedented scarcity of water, even when this was coupled with most unseasonable cold and fogs in September, which considerably diminished the output of cotton. Lord Cromer had indeed good reason to write in 1901 :

> 'Had it not been for the labours of the eminent hydraulic engineers, who for the last seventeen years have placed their services at the disposal of the Egyptian Government, the most skilful financial assistance would not have availed both to place the Egyptian Treasury in a position of assured solvency and to meet in any adequate degree the constant demands which are the necessary accompaniment of a policy of reform.'

Such are the outlines of the long history of the Barrage, designed by Frenchmen and brought to perfection by Englishmen. Both nations can share in the credit of the work, and it is pleasant to record once more the generous and graceful act by which the chief of the English engineers recognised and acknowledged the merits of his predecessor. Sir Colin Scott Moncrieff discovered Mougel Bey living in obscurity and oblivion, weighed down by poverty and neglect. It was owing to his intercession that the poor old man was rescued from want, and, by means of a pension granted by the Egyptian Government, enabled to spend his remaining days in comfort and honour. Both nations are entitled to be proud of this act of poetic justice, which added a lustre of its own to the glory of the completed Barrage.

CHAPTER VI

THE CORVÉE

From time immemorial the peasantry of Egypt have been liable to the corvée in some form or other. In a country depending for its existence upon the proper maintenance of its dykes, it was only natural that the whole population should turn out to perform the necessary work. But a useful custom very easily degenerated into a galling slavery under Oriental despotism. Rulers with absolute power of life and death over their subjects, who regarded the land they ruled as their own personal property, could not be expected to make much distinction between works carried out for the general good and those designed merely for their own convenience and aggrandisement. The Pyramids and others of the mighty remains of ancient Egypt stand as monuments of the greatness of the Pharaohs, but no less of the miseries of countless generations under the system of forced labour, which is known in our time as the corvée, a term applied sometimes to the forced labour itself, and sometimes to those who perform it.

In the early chapters of the Book of Exodus there is a brief but pregnant description of the sufferings of the Israelites when subjected to the burden.

'And the Egyptians made the children of Israel to serve with rigour:

'And they made their lives bitter with hard bondage, in morter, and in brick, and in all manner of service in the field: all their service, wherein they made them serve, was with rigour.'

'And Pharaoh commanded the taskmasters of the people, and their officers, saying, Ye shall no more give the people straw to make brick, as heretofore: let them go and gather straw for themselves. And the tale of the bricks, which they did make heretofore, ye shall lay upon them; ye shall not diminish ought thereof. Let there more work be laid upon the men, that they may labour therein.'

'And the taskmasters hasted them, saying, Fulfil your works, your daily tasks, as when there was straw. And the officers of the children of Israel, which Pharaoh's taskmasters had set over them, were beaten, and demanded, Wherefore have ye not fulfilled your task in making bricks both yesterday and to-day, as heretofore?'

'And the officers of the children of Israel did see that they were in evil case.'

'And the Lord said, I have also heard the groaning of the children of Israel, whom the Egyptians keep in bondage.'

'But the children of Israel hearkened not unto Moses for anguish of spirit, and for cruel bondage.'

Doubtless the amount of forced labour varied from time to time, according to the ambition or the caprice of the rulers of the country. But the annual necessity for watching the dykes during the flood, and repairing them when it was over, never permitted the custom to fall into disuse, and the knowledge that this great instrument was always ready to hand must have been a powerful incentive to any King or Caliph who wished to send his name down to history as the author of a mighty work. No wonder that after so many centuries of practice the Egyptians are the most patient and efficient spade-workers in the world. The wretched peasantry of Egypt must have blessed the accession of the undistinguished Sovereigns who had no desire to add to their fame either by building at home or by conquest abroad. To them the glories of a Rameses or an Amenemhat must have been small compensation for their 'anguish of spirit and cruel bondage.'

So long as basin irrigation continued to be universal, there was much to be said on behalf of the corvée, if the system was justly and impartially administered. During the months of the flood, and those preceding it, when the land was lying dry and baked, there was little or nothing for the agricultural population to do except to clear the shallow flood canals, repair the dykes, and protect the river-banks. If the labour was compulsory, it was, at any rate, everybody's interest to perform it. The work lay at their own door; they were not dragged away to a distant province. There must always have been abuses in practice. The humbler folk did more than their share, and so on, but in theory it was not bad.

All this was changed with the introduction of perennial irrigation and the digging of the summer canals. Owing to their depth, and sometimes, also, to their faulty construction, the silt deposits in them were very great, and the whole corvée was called out to clear them, though very few were interested in them. More than that, a man's own home was no longer the scene of his labours. The labourers of one province were called in to work in another. Each year the corvée worked from January 15 to July 15, clearing the canals and repairing the banks. From August 1 to November 1 they guarded the Nile banks in the flood. Every year an extensive programme of work was sketched out, but before it was finished they had to hurry off to flood-protection duty. Not only unpaid, they had to find their own tools, and provide their own commissariat, a double hardship on men out of their own neighbourhood. During the flood, when they lived in booths built for themselves on the Nile banks, they had to find their own lanterns, and even, like the Israelites of old, their own straw and brushwood, to save the dykes from the action of the waves. New works were carried out in the same fashion.

Nor was this the sum of their grievances. The increase of summer irrigation made the months immediately preceding the flood a very busy instead of a very slack time of year. The value of the cotton crop made everyone most anxious to secure it, so the larger proprietors kept their tenants at home for that purpose. The numbers of the corvée decreased, and the burden of it fell more and more upon the poor fellaheen. Ministers and high officials from the Khedive downwards employed the corvée to work on their own private estates. Other persons, influential by station or by bribery, secured the like advantage, and robbed their humble neighbours of their labour, their last remaining possession. Under Said Pasha the corvée dug the Suez Canal. Under Ismail they dug the Ibrahimiyah Canal, the sole object of which was to benefit his private estates in Upper Egypt. Even the splendid carriage-road that runs from Cairo to the Pyramids of Gizeh was raised on the same foundations that the Empress Eugénie might travel there in comfort after the opening of the Suez Canal.

Only a nation inured to slavery could have endured it. Many, indeed, labouring under the burning sun, unpaid, unfed, and unclothed, succumbed. But what did that matter when other human beasts of burden were there to take their place? The activity of the survivors was kept up by the whip, the traditional motive-power in Egypt. Nominally, all between the ages of fifteen and fifty were liable to serve, but there were a multitude of exceptions, including teachers, holy men, students, certain classes of tradesmen, and others. The law of 1881 laid down that anyone might exempt himself by providing a substitute or by paying a cash ransom. But as there was no penalty imposed for not paying, every man of any position freed himself from the obligation without paying the tax, and the whole burden of the corvée fell more than ever on the poorer classes. The régime of the kurbash, or whip, and the grosser abuses of the system, vanished immediately upon the English occupation, but all the earthwork maintenance was still performed by this unpaid labour.

It was, as I have said, the spectacle of the dumb misery of the fellaheen that particularly stimulated the English engineers in their task of repairing the Barrage. The first relief came in

1884, when the Nile was held up to a higher level at the Barrage. This had a twofold effect, for the canals did not require to be cleared to so great depth ; and the higher level of the water enabled them to be laid on a better slope, which diminished the deposit of silt. In fact, the partial use of the Barrage in 1884 reduced the amount of silt deposit by 26 per cent. In 1885 the first step was also taken in a new direction : £30,000 were advanced towards the experiment of clearing some canals in the provinces of Menoufia and Gharbiah by contract. As usual, the gloomiest forebodings were uttered on every side. It was said the fellaheen would not work voluntarily. The whip alone was the only stimulus to which they were sensible. The experiment was bound to fail.

Once more the croakers were wrong. The fellah justified the confidence of Mr. Willcocks and his colleagues, that he was not unlike other men, and would work gladly for a wage. And so the struggle for emancipation went on with increased vigour. In 1886 £250,000 were provided by the Caisse for the reduction of the corvée, so that while in 1883 107,000 men had been called out in the Delta alone, in 1887 the number had been reduced to 27,500. Finally, December, 1889, saw the last of the system, and the performance of earthwork maintenance by the corvée was finally abolished. The Government supplemented the £250,000 a year received from the Caisse by another £150,000. In former days the labour required to clear the canals was estimated in the Delta alone at £530,000. For the sum of £400,000 Egypt got rid of the burden throughout the whole country. It was a bargain well worth making at a far higher cost. No greater boon could have been conferred upon the fellaheen. No longer are their lives made 'bitter with hard bondage, in morter, and in brick, and in all manner of service in the field.' That 'evil case' is ended. They labour indeed, but it is voluntary labour, without 'groaning and anguish of spirit.'

The Government, of course, retains the right to call out the corvée in case of any grave national emergency, and every year, too, it is called out to guard the banks in flood-time. But the flood-corvée is cheerfully borne. It entails no hardship on the people. Its incidence, too, has been very much diminished, as the following figures show (the period of service is 150 days):

Year.	*Number of Men called out.*	*State of River.*
1888	59,000	Low flood.
1889	50,000	Medium flood.
1890	48,000	,,
1891	45,000	,,
1892	84,000	Very high flood.
1893	33,000	Low flood.
1894	49,000	High flood.
1895	37,000	Medium flood.
1896	26,000	,,
1897	11,000	Low flood.
1898	19,000	High flood.
1899	8,000	Very low flood.
1900	14,000	Low flood.
1901	9,000	,,
1902	5,000	Very low flood.

The figures tell their own tale. Experience and good organization have enormously decreased the number of men called out, so that it is now but a very slight burden upon the people. The date of calling out has also been altered to August 15 instead of August 1. The possession of the Soudan, and an accurate record of the state of the gauges in those regions, will also assist in making more accurate forecasts of the nature of the flood. In 1900 more men were called out than were required, because the flood came down early and promised to be a high one, but failed to fulfil expectations. The levels in the Soudan were known, but, there being no previous experience to judge them by, no inferences could be drawn from them. But the Soudan readings will be more and more useful as time goes on.

But even apart from this, there is great hope of a steady diminution in the numbers called out, and even that, except in years of high flood, the corvée may not be required at all. The record of the year 1901 is of remarkable promise in this respect. North of Cairo no corvée was called out at all, for the first time in the history of Egypt. The flood was low, but not exceptionally low. In any year Upper Egypt is responsible for far the greater number, and this is largely due to the extent of the basin banks which have to be guarded. It is remarkable that the two districts in which the greatest number were called out were in charge of native Egyptian inspectors, who were no doubt influenced by the old tradition that vast numbers of men should be employed. At any rate, as perennial irrigation increases in Upper Egypt, fewer men will be required, concurrently with the disuse of the basin dykes. And it seems likely that in years to come the whole task will be performed by contract labour ; though the power of calling out the corvée will always be held in reserve in case of any specially dangerous flood. But in such a case the difficulty will be rather to prevent the work from being hindered by excessive numbers. There will be no doubt of the willingness to serve of practically the whole population.

CHAPTER VII

RESERVOIR PRELIMINARIES

In dealing with the history of the Barrage I have somewhat anticipated the order of events. It was the prospect of the coming Reservoir and an increased water-supply in summer that urged on the engineers to make assurance doubly sure by placing the strength of that structure beyond all doubt. It is time to pass from the Delta to Assouan.

No country in the world tells its story more readily to the traveller than Upper Eygpt. As he passes up the broad waterway of the Nile, he may survey the whole life of the land without stirring from the deck of his steamer. If he has been in Egypt before, he cannot fail to be struck by the growth of its prosperity; it forces itself upon him in the bearing of the people, and in the number of their flocks and herds, now as ever the outward and visible sign of material well-being. Even the squalid clusters of mud huts, often roofless, or covered only by a few loose canes, dirty and miserable as they seem to Western eyes, with nothing substantial among them except the tomb of some sheikh or the inevitable pigeon-houses, are only proofs of the genial climate, which makes a roof overhead, and clothing as well, among the least of the necessities of life. The people themselves, hard workers as they are, have a happy and prosperous aspect, and the crowds of naked children, brown as the waters by whose edge they play, look as cheerful and contented as the vast colonies of pigeons, which live under very similar conditions to their owners.

On each side of the river-valley, here and there, especially on the eastern side, coming right to the water's edge, rise the hills of the desert. Where the domain of the water ceases a man may stand with one foot in the bare and barren sand and the other in the most fertile soil in the world. Everywhere along the bank, hour after hour, day after day, the traveller may see the peasants lifting the water with the primitive shadoofs, tier upon tier, up to the level of the fields, or the oxen turning the sakieh. A hundred times a day he will have borne in upon him the fact that all he sees, from the kid upon the dykes to those obelisks of modern Egypt, the tall chimneys of the sugar factories, owes its existence absolutely to the water. Close behind the teeming villages and the luxuriant crops, the palm-trees and the acacias, the sugar-canes and the maize-fields, rise the gaunt limestone rocks and the sandy desert, fit emblems of the famine that is ever ready to swoop down should the water fail.

Even as late as December, steaming up the 550 miles of river, often half a mile or more across, between Cairo and Assouan, against the strong current, watching the majestic sweep of those wide waters pouring irresistibly towards the sea, it is hard to realize the anxiety of later months. But May or June has a very different tale to tell. To take and store the precious water, which now during the flood and winter rushes down in untold volume, to be lost and squandered in the sea, and use it to feed the lean summer months, is almost absurdly obvious. No wonder that since Mehemet Ali gave so great a stimulus to the cultivation of cotton and sugar the idea of the Reservoir has been constantly in the minds of the rulers of Egypt. The strange thing is that so many hundreds, even thousands, of years should have elapsed without any attempt of the kind being made. Perhaps it was due partly to the reverence felt for the mighty and inscrutable power of the great river, partly to the passive fatalism innate in the Oriental mind. A few years ago Sir Benjamin Baker asked a prominent and representative land-owner in Egypt, a Pasha, and a descendant of the Prophet, what he thought of the prospects of a Nile Reservoir. With a shrug of his shoulders, he replied that 'if it had been possible it would have been done 4,000 years ago.'

He reasoned wiser than he knew. At the second cataract above Wadi Halfa there are marks upon the rocks and other indications which go to show that a Dam once existed at that point, used to regulate the flow of the Nile. Swept away by some unrecorded disaster, no other direct knowledge of it remains. But it is far from unlikely that Herodotus, in his account of Lake Mœris in the Fayoum, has mixed up some tradition of this ancient work in Nubia. At any rate, whatever be the truth about Lake Mœris, his account proves beyond all question that the idea of a Reservoir was familiar to the ancient Egyptians.

The tradition of a Reservoir somewhere on the upper waters of the Nile lingered long in Egypt. There is a curious reference to it in a book of travel by F. Vansleb, a Dutchman who visited Egypt about the year 1670. The fertility of the Nile flood is caused, he says, by a fall of dew, which usually takes place on June 17, just after the appearance of the 'green' water. This dew purifies the foul water, and makes it swell by fermentation. 'Some of the country,' however, he proceeds, 'that are ignorant of the true causes of this increase, imagine that it proceeds from a large pond in Ethiopia *in the river itself*, which the Abyssins begin to open about June 12, and let the water out by degrees, more and more till September 14, by which time they begin to shut it again. But this is a foolish fancy of the Copties.'

We have seen how the tradition of Lake Mœris fascinated Mehemet Ali; but the methods of Haroun-al-Raschid were not suited to solid engineering works, as the history of the Barrage too plainly shows. None of his descendants, with the exception of the Khedive Ismail, had the wit to conceive or the ability to execute such an undertaking, and Ismail's fantastic imagination was fully occupied in other directions. Fortunately for Egypt, the project had to wait until the success of the Barrage made the time ripe for its execution, and until skilful brains and strong hands were ready to plan and carry it out in the most efficient manner possible.

There were three problems to be faced: first, Where could such a Reservoir be erected? second, What arrangements could be devised to avoid the danger of large silt deposits, which would soon seriously diminish the capacity of the Reservoir, and, if allowed to accumulate, render it in no long time entirely useless? third, Supposing that the difficulties of site and design could be overcome, where was the money to be found? During the first years of the British occupation, while Egypt was still painfully struggling upwards from the abyss of bankruptcy into which she had been cast by the mad whirlwind of extravagance in Ismail's reign, it was no time for the inception of original works on a grand scale. But in 1890 the matter became an affair of practical politics, and was at last seriously taken in hand. Meantime discussion had been raging as to the best locality for the Reservoir. An American gentleman, Mr. Cope Whitehouse, took up the case of the Wadi Rayan, a depression in the desert to the south-west of the Fayoum. This, he maintained, was the real site of the ancient Lake Mœris, and here the Reservoir ought to be. He had no professional knowledge, and he was utterly wrong in his ideas; but his vehement method of controversy kept the subject thoroughly alive. The whole land was filled with his clamour, and every expert was forced to give his own views in self-defence. The debate served a useful purpose. Gradually it came to be recognised that the river-bed itself was the proper place for storing the water by means of a Dam. Authorities differ as to whom belongs the credit of first making this suggestion, or, rather, of first reviving the tradition of the past; but it seems pretty clear that Sir Samuel Baker suggested the construction of a Dam at the first cataract at Assouan as far back as 1867.

However that may be, in 1890 the Government took the matter up, and charged Mr. W. Willcocks with the task of examining the river north of Wadi Halfa, reporting upon the best available site for the Dam, and preparing a design for it. After a careful survey, his plans were completed in 1894, and his design for a Reservoir at Assouan was then submitted to an International Committee of Engineers, consisting of Sir Benjamin Baker, M. Boulé, and Signor Torricelli. Mr. Willcocks' plans were, with some modifications, accepted by a majority of the Commission, and to him belongs the honour of having designed the Dam. The selection of the Assouan site solved the first of the three difficulties. There is at this point an extensive outcrop of granite clean across the valley of the Nile, which it was thought would give sound rock everywhere at a very convenient level for the foundations of the Dam. Moreover, the trough of the river above the cataract and a long way south of it is exceptionally deep, and this makes it possible for a greater amount of water to be stored up behind the Dam. But the prime necessity was for a solid foundation. Elsewhere in Egypt the bed of the Nile is composed of shifting sands, on which it would have been impossible to build a Dam capable of holding up so great a head of water.

Mr. Willcocks' design solved the second difficulty, the problem of constructing a Dam strong enough for the purpose, and yet of avoiding the danger of filling up the Reservoir by too great accumulations of silt. The other great Dams in the world, as, for instance, that built by Sir Arthur Cotton on the Godavery River, in India, are solid throughout. They are planned so that the rising flood shall pass freely over the top of them. But the Assouan Dam is of a type previously unknown, and its success ought to stimulate perennial irrigation in many parts of the world where such projects have hitherto proved failures. Its principle is that even the highest flood shall pass, not over it, but through it. To this end it is pierced with 180 openings, which are like tunnels in the great mass of masonry. The openings are controlled by powerful sliding-gates worked from above. During the months of the flood every gate will be up, and the 'red' water, carrying all its heavy burden of silt, will pass through without impediment. Later in the year, about the end of November, when the flood has subsided and very much less matter is carried in suspension, the sluice-gates begin to be gradually closed, and by the end of February the Reservoir is full, without having affected the normal discharge of the river in any appreciable degree. From April to July the water thus stored up is let out by degrees for employment, according to the state of the river and the requirement of the crops. By the time the next flood begins to come down all the stored water will have passed out, and every sluice will be once more open to give free passage to the rising stream. Although the Nile in December and January carries an insignificant amount of sediment compared to that brought down in August and September, it yet brings down a very considerable quantity, far greater than most other rivers at any time, and quite enough to go a long way towards silting up the bed of the Reservoir, if it was allowed to remain. But for this the river provides its own remedy: every year the force of the flood will act like a gigantic broom, sweeping the floor of the Reservoir. The sluices, arranged in sets of five, are distributed at different levels, according to the formation of the river-bed on the upstream side, so as to facilitate this process to the utmost. During the months of the inundation the Nile at Assouan pours down for weeks together a volume of 10,000 tons of water per second, and sometimes as much as 14,000 or 15,000 tons per second. The rush of this stupendous mass is sufficient to assure us that there will be no silting up of the Reservoir.

Two of the difficulties had been thus overcome, when, from a new and unexpected quarter, a storm sprang up, which very nearly brought to a standstill the rising fabric of Egyptian prosperity. The project of the Reservoir would have raised the level of the water, and held up the river above the Dam to a head of 100 feet; this would have involved the temporary submersion every year of the island of Philæ, with its famous Temple of Isis, Pharaoh's Bed, and other monuments. A terrific hubbub arose. Archæological and antiquarian societies, which until then had sometimes belittled the monuments of Philæ as belonging to an inferior period, poured in their protests. People who had never heard of Philæ before, but who were none the less influential for that, joined in the outcry. Diplomatists, whose one desire was to embarrass our progress in Egypt, took up the cause of Art with a will. These champions of humanity at large forgot the poor fellaheen, to whom the extra water means all the difference between misery and happiness; nothing would satisfy them but the complete abandonment of the project. The engineers fought stoutly in the interests of Egypt; they offered to raise the whole of the monuments bodily, or to transport them to the neighbouring island of Bigeh; but, though they saved the Dam, the original design was lost, and the Dam to-day is 33 feet lower than it ought to have been. The foundations of Philæ have been underpinned and strengthened, the island will only be partially submerged, and the injury to Egypt can only be faintly estimated.

Was the sacrifice worth it? The value of Philæ lies in its beauty more than in its antiquarian interest. No one who has witnessed night after night the glorious sunsets on the Nile, the mysterious charm of the changing waters, the dark belt of palms reflected in the river below and standing out in strong contrast against the sky, the limestone cliffs of the desert clear-cut in the dry air, and flushing pink in the radiance of the indescribable after-glow, no one who has seen the Temples of Karnak, could hesitate to make so small a sacrifice, in comparison, for the sake of the riverside people. Moreover, Egypt is rich in treasures of the past, as yet undiscovered, and wanting only money for their development, which the Reservoir would in time supply. And how few people visit Philæ at all! Surely, even in a country a thousand times poorer than Egypt in artistic and archæological interests, the well-being of the living and of the unborn should have prevailed.

If all those who joined to swell the uproar had been really disinterested lovers of the beautiful, there would have been small reason to complain of their insistence. Enthusiasts can hardly be expected to listen to the voice of reason, and Philæ has charms to soften the heart of the most savage utilitarian. Its fate is a mournful necessity, but it is a necessity, for the question of the Reservoir had come to be a question of existence for Egypt. Even the advantage gained by the opposition in lowering the height of the Dam is only a delay. On the pylon of the Temple of Isis at Philæ is carved a huge representation of the Pharaoh of the day, one of the most degenerate of the Ptolemies, catching his defeated enemies by the hair of their heads with one hand, an uplifted sword in the other. The whole is a copy of the work of his warlike ancestors, and even as a copy it is a delusion and a sham; for he won no victories, defeated no enemies, and, indeed, scarcely ventured outside the walls of his harem. The apparent victory of these lovers of Philæ, to call them by their more honourable title, was not less delusive. Philæ is doomed. Between half drowned and wholly drowned there is not much difference in the case of an island, certainly not a difference worth fighting for, and the Dam will be raised to its full height, perhaps as soon as Egypt is ready for the extra water.

The financial difficulty remained. In spite of the prosperity of Egypt, she is, as everyone acquainted with her history is aware, bound hand and foot by international fetters in matters financial. The Caisse de la Dette, founded to protect Egyptian creditors against the dangers of bad administration, has remained to be an obstacle to any improvements that must benefit these interests. The practical outcome of the system is that, if the Caisse be hostile—and hostile it has often been—no public work like the Nile Reservoir can be carried out without the imposition of extra taxation to the amount of double the annual expenditure required.

Time passed, the need became more pressing, but the prospects of the Reservoir seemed further off than ever. Besides the regulation of her water-supply, Egypt had on her hands the question of the Soudan. From every point of view the re-conquest of that province and the upper waters of the Nile was a prime necessity; no one could tell how long the war might last, or how great the expense might be. It seemed impossible that she could bear the cost of two such enterprises simultaneously, and under such circumstances her credit would not have been sufficient to raise the capital sum required on anything like reasonable terms. Not only so, but by the peculiar constitution of Egyptian finance it was illegal for her to raise a loan without the consent of the Caisse, a consent which it was impossible to obtain.

But, fortunately for Egypt, there were a few men with clearer vision and more faith in the future, and chief among these was Lord Cromer. The statesman who had controlled the tangled destinies of Egypt through so many dark years, and baffled so many tortuous intriguers, as well as more open foes, was not the man to despair in such a situation. In 1897 the first negotiations were quietly opened with Sir E. Cassel. Then came the vote of the majority of the Caisse to grant £500,000 towards the Soudan Railway, and the successful action taken in the Courts against that vote. Everyone knows how this seeming defeat was turned to overwhelming victory by the decision of the English Government to grant £750,000 for the railway on certain conditions. The enemies of England in Egypt received a staggering blow.

But the story was not yet complete. In April, 1898, Sir E. Cassel arrived at Cairo; in one day the details of the arrangements to finance the Dam were settled; all that night the lawyers drafted the necessary documents; a Council of Ministers was hastily called in the morning, and the contracts were signed. Sir E. Cassel was to provide the necessary funds for the execution of the work, £2,000,000; repayment by the Egyptian Government was to be deferred altogether for five years, and then to be spread over a period of thirty years. The first payment of about £78,000 is included in the Budget for 1903.

Looking back now after five years of prosperity, when Egyptian securities have actually increased in value, while Consols themselves have so greatly declined, it is easy to see that the statesman and the financier were justified in their faith. But in those days it needed a clear vision and a stout heart to calculate thirty years ahead—nay, even five—in a country so much the sport of international politics. The Soudan Campaign was not yet ended; behind the dervishes there loomed vague possibilities of worse complications. The Egyptian Government made a good bargain then, though it would doubtless make a better now. But it was then, and not now, that the business had to be settled.

CHAPTER VIII

THE DAM AND THE NEW BARRAGES

ONCE the financial difficulty was settled, no time was lost in setting to work. As soon as the flood of 1898 began to subside, Messrs. Aird and Co., the contractors, were busy with the foundations of the Dam. Five years was the period allowed by the contract, but a succession of low Niles gave unusual facilities for the work, and everything was completed before the flood of 1902, a year before the specified time.

From its vast proportions, the Dam is infinitely more impressive to the imagination than any other of the irrigation works in Egypt. But from an engineering point of view its construction was a plain, straightforward business compared with the difficulties of building a Barrage, where the river-bed offered no more solid foundation than shifting sands. Still, there was a moment, on the first uncovering of the river-bed, when its fate seemed to hang in the balance. The Assouan site had been selected principally because the outcrop of granite, there running clean across the valley, would give, it was thought, solid foundation at a convenient level. It was found that in some places the rock was rotten to a depth of 40 feet.

It was an anxious moment, both from an engineering and a financial point of view. Every foot of rotten rock meant a considerable addition to the calculated expense, besides modifying the building plan. Once more Lord Cromer's strong will saved the situation. On the financial side he stood on firm ground, and he proved as good an engineer as he had been a financier. Solid rock was reached, and the work went steadily forward. Ten thousand men was the usual sum of those employed, and of these 800 were Italian stone-cutters specially brought over to deal with the tough granite of which the Dam is built. Granite and Portland cement are the two great materials used for welding the fetters of the Nile.

The Dam is about one mile and a quarter in length, and at its deepest point it is 126 feet high. Sixty-five feet of water can be held up when the reservoir is full, and it is capable of storing about 1,200,000,000 cubic metres of water—that is to say, about the same amount of water as passes through Assouan in a single day when the flood is at its height. The face of the wall is a slope on the downstream side, and its width at the bottom corresponds approximately to its height. Seven hundred and eighty thousand cubic yards of masonry have been used. On the western side a ladder of four locks gives passage to boat and steamer traffic at all seasons.

All parts of Egypt are to benefit in a greater or less degree from the extra summer supply. The original calculation assumed an amount to be distributed of 1,065,000,000 cubic metres. It was allotted as follows :

	Cubic Metres.
South of Assiout	170,000,000
Assiout to Cairo with the Fayoum ...	510,000,000
Gizeh Province	85,000,000
Lower Egypt	300,000,000

This division meant that 52,000 acres could be reclaimed to cultivation in the Fayoum, and 120,000 acres in the Delta. Further, south of Assiout 200,000 acres could be converted to perennial irrigation by means of pumps upon the Nile banks. In Middle Egypt 458,000 acres could be converted to perennial irrigation, and in Gizeh Province 106,000.

A low Nile on the average comes about once in five years, but assuming that it happened every year, these results may be expressed in terms of money on the basis that conversion to perennial irrigation increases the yield per acre by £2 annually, and that reclaimed land produces a yield valued at £5 annually. This gives an increase of value in the annual yield :

	£E
South of Assiout	420,000
Middle Egypt and Fayoum	1,176,000
Gizeh Province	212,000
Lower Egypt	600,000
Total ...	£E2,408,000

while the direct annual gain to the State Exchequer in rental and taxation would amount to £E378,400.

This does not exhaust the full extent of the benefits of the Reservoir on which a money value can be placed. For some time Egypt had been living beyond her real resources in the matter of water. Encouraged by a series of good years, the acreage of cotton had been greatly extended. The cotton-plant is very hardy, and can retain its vitality for a certain period on a very scanty and irregular supply of water. Cultivators, especially in the Delta, had taken advantage of this quality up to the very hilt, and the annual crop had reached an amount of 6,000,000 kantars.

Taking the not very high price of 175 piastres per kantar (1 kantar = nearly 100 pounds), the annual value reached £10,000,000. A season like that of 1889, when the summer supply was very low, and not the lowest on record, would mean the loss of at least one-tenth of this amount, and even more, in spite of the most successful working of the Barrage, and the most careful system of rotations. Against such a loss the Reservoir is a complete insurance, and, as a low year cannot safely be reckoned as occurring less than once in five years, the annual value of such insurance must be set down as £200,000. The figures give some idea of the value of the new supply. The estimate does not err on the side of exaggeration. It was framed in the most cautious and conservative manner possible, and, in fact, it would be by no means rash to put the total annual value a good deal higher.

According to the financial arrangement, the first payment towards defraying the cost of the works was included in the budget for 1903, but no fresh taxation for the purpose was to be imposed till 1904, and even then the full amount of direct benefit to the Exchequer will not be realized till 1910. Time is thus given for the full effects of the change to be felt, and the Reservoir will be paid for out of its own profits. The alteration of the basin lands to the new system must take some time, and their cultivators will thus be given full opportunity to familiarize themselves with the new methods of agriculture which they will have to employ.

The scheme of distribution allotted nearly 50 per cent. of the Reservoir supply to the Fayoum and the province between Assiout and Cairo. Just south of Assiout is the head of the great Ibrahimiyah Canal, which not only supplies these provinces, but also feeds the Bahr Yusuf, which waters the Fayoum. It was necessary, in view of the increased discharge, to widen the upper reaches of this canal, and to provide it with a new regulating head. But more than this was required to insure its receiving the proper proportion of water whenever the sluice-gates of the Dam were opened. Accordingly, at the same time that the foundations of the Dam at Assouan were laid, a new Barrage was begun at Assiout just downstream of the head of the Ibrahimiyah Canal.

In principle the Assiout Barrage is exactly the same as that at the point of the Delta, and the difficulties of construction were also exactly similar, for in both cases the foundations had to be laid on the same shifting sands, and as each section of the work was undertaken a portion of the river had to be diverted from its course by means of temporary earthen dams. The shifting nature of the river-bed, the almost personal malignity of the water, constantly bursting through in countless springs, each of which had to be separately dealt with and suppressed, called forth the highest exercise of engineering skill. But the experience gained in the long struggle with the imperfections of the earlier Barrage infallibly told its tale. Every difficulty was successfully encountered, and the Assiout Barrage was completed by the summer of 1902, and was able to hold up the 10 feet of water required of it, without any failure, at the first attempt.

The visible part of this Barrage, which is just over half a mile in length, consists of a viaduct or bridge with 111 archways, each 16 feet 5 inches in width, closed by strong iron gates, working in grooves made in the supporting piers, and raised or lowered from above. In contrast to the Delta Barrage, it is built throughout of stone, and not of brick. At the western end is a lock, the largest in Egypt, through which the largest of the boats that ply upon the Nile can easily pass. Below the water lies the strength of the structure. The viaduct rests upon a solid platform of granite and cement, 10 feet deep and 87 feet wide, set at a suitable depth below the bed of the river. As a further precaution against the action of the water, there are also below the platform two continuous lines of iron sheet-piles, with hermetically sealed joints. With such a series of obstacles to encounter, the danger of the water forcing its way through underneath the Barrage is small indeed. The whole amount of masonry used in this Barrage is 220,000 cubic yards.

The prosecution of these great works in Upper Egypt by no means exhausted the activity of the Irrigation Department; indeed, it would almost seem that the building of Barrages has become part of its ordinary routine, for a third remains to be chronicled. This is the Zifta Barrage on the Damietta branch of the Nile, halfway between the point of the Delta and the sea. Because it lies in a district unvisited by tourists, though very important commercially, its construction has not been heralded by any blowing of trumpets. Yet it is a work of the very first class, and deserves to be reckoned among the greatest of the triumphs of the department. Built on the same plan as the Assiout Barrage, and, like it, capable of holding up 10 feet of water, it is designed to secure a better distribution of the supply north of the Delta Barrage. As the area of the cultivated land extended gradually northward, it became apparent that the canal system taking off from above the original Barrage was becoming too long to admit of the water in times of pressure reaching the northernmost parts of the country. There were some who held that a second Barrage, with a new system of canals taking off from it, should have been erected on the Damietta branch, even in preference to the new stone weirs, which have increased the strength and distributing power of the old one. The dispute has been happily settled by the adoption of both projects, and with the Zifta Barrage completed in time for the summer of 1903, and the new supply from Assouan, the reclamation of the northern lands will go steadily forward.

It has been already pointed out that it is almost as important to get the water off the land as to get it on, and the proper drainage of the Delta lands has been the necessary complement of all the new schemes. Though eclipsed by the splendour of the Reservoir and other creations more taking to the eye, the performance of the engineers in this direction during the last few years has been sufficient to make them very memorable in irrigation annals. Since 1896 about 1,000 kilometres of new drains have been dug, and nearly as great a length of existing drains remodelled. It has cost the Egyptian Government close upon a million of money, but that this expenditure has not been thrown away is proved by the great rise in the value of all the lands affected.

The Zifta Barrage cost about £500,000, the Assouan Dam and the Assiout Barrage £3,200,000. Apart from the ordinary expenditure on maintenance and the suppression of the corvée, the twenty years ended 1902 have seen an expenditure of £9,000,000 devoted to irrigation and drainage. There could be no greater proof of the wisdom of those who have directed Egyptian policy during that period. However pressed they have been at different times by demands for immediate expenditure on other objects when resources were low, they have always adhered steadily to a policy of liberality towards public works likely to be of a remunerative character. In no other country do economic laws work out their results with greater directness and certainty. Reproductive expenditure is really worthy of its name, and brings its visible and tangible fruit almost without a moment's delay. As they have sown, so have they reaped. There can be no comparison between this expenditure and the benefits it has conferred upon Egypt. Three salient examples may be given to point the force of these remarks.

First, the works undertaken by Colonel Ross to improve the system of basin irrigation in Upper Egypt. 1877 was a year of very low flood, and nearly 1,000,000 acres were sharaki—that is, entirely exempted from taxation owing to absence of irrigation. In 1899, a worse year—in fact, the lowest flood of the century—the sharaki lands were no more than 250,000 acres, a result mainly due to the successful reforms carried out by Colonel Ross in 1888-89. 1902 was a year very similar to 1899, and the sharaki acreage was no more than 140,000 acres.

Second, the case of the Assiout Barrage. The work was finished a year before the time named in the contract. On August 15, 1902, the usual date of filling the basins, the flood was exceptionally low, and it was decided to lower the gates of the Barrage. By so doing the water-level of the canals was raised by 1½ metres, an increase which was more than sufficient to avert the threatened disaster. The money value of the crops thus secured to the land-owners of the Fayoum and Middle Egypt is estimated at not less than £600,000. The cost of the new works at Assiout, including the new regulator on the Ibrahimiyah Canal, was about £875,000. Thus, in the first year of their existence they nearly repaid their whole cost, and that, too, as it were, by a side-wind; for the Barrage's real function is to hold up the river in the low summer season, and not in the flood.

Third, the case of the Assouan Reservoir itself. The summer supply in 1903 has been the lowest on record; the discharge at Assouan has fallen to

200 cubic metres per second. By means of the Reservoir this supply has been actually doubled. Had the flood been late in coming down, it would have been impossible to distribute the water at so liberal a rate. The Soudan gauges gave warning that an early flood was to be expected, and thus the authorities were able to calculate with certainty, and open the sluice-gates with much greater freedom. But the calculation would have been a useless exercise unless there had been a store to draw upon. At the lowest computation the loss avoided may be reckoned at a couple of millions.

The administrators of Egypt have had many difficulties and obstacles in their path, but to be able to point to such results is a great compensation ; their efforts need no formal monument.

CHAPTER IX

THE INAUGURATION OF THE RESERVOIR

DECEMBER 10, 1902, was the official date of the inauguration of the Reservoir, a memorable day in the history of Egypt, and worthy to be marked with red even in the unchanging Mohammedan calendar.

The making of the Dam has been a great time for Assouan. The town has thriven and prospered beyond all knowledge since the days—not so very long ago—when two British battalions occupied the barracks on the hill overlooking the river to the south. The barracks are crumbling to pieces now, and only one or two blockhouses remain as memorials of the past state of siege and fear of dervish raids. The tide of war has rolled far away and spent itself utterly in remote corners. The whole of the Soudan lies between Assouan and the frontier of any possible enemy. Even the yellow fort is untenanted save by a few policemen. True, for four years an army of 10,000 men has been marshalled here, but it was an invasion of the arts of peace, creative and not destructive. Possibly the inhabitants would have liked their occupation to go on for ever; but even the best of times must have an end, and it was a good occasion for a holiday. In every Nile village flags and bunting were flying. For once the fields were deserted, and everywhere the people crowded to the bank in the hope of catching a glimpse of the Khedive and his distinguished guests. Here and there the gaffirs, or local policemen, lined the shore, standing stiffly to attention, or saluting with their Remingtons in the regulation attitude of European soldiers, which contrasted quaintly with their loose, flowing robes and white turbans.

At Assouan the faithful subjects of the Khedive surpassed themselves. Sunny Assouan lends itself readily to a festal garb. Situated where the Nile broadens out after emerging from the rocky defile of the cataract, the town has a most picturesque aspect at all times; its embanked river-front makes it the neatest of all the cities of Upper Egypt. Elephantine and the Sirdar's Island rise green and smiling out of the broad bosom of the river; the perpetual blue sky makes everything doubly attractive to the Northern visitor. Dressed for the festival, it was a charming scene. Triumphal arches of the sacred yellow and brown, gorgeous hangings and many-coloured festoons, bore testimony to the Oriental love of vivid hues; steamers and dahabiehs, moored in line along the shore as well as along Elephantine Island, vied with each other in their decorations, and numerous feluccas were plying to and fro, half hidden by their burdens of flags and palm-leaves. At night thousands of lamps decorated shore and river alike, and the whole scene resembled nothing so much as a Henley in Regatta Week, with its illuminations unrestrained by doubts of weather. To the ear, however, the voices of the night told a very different tale. The crooning song of the Nubian boatmen, 'Great is the Prophet, praise be to him!' accompanying the creaking of their clumsy oars with monotonous persistency, sounded weird and barbaric over the twinkling waters. The bustle in the town, too, was no mere ordinary festal murmur; for this was the month of Ramadan, and the feast of lamps meant a great deal to all faithful Moslems. All day long they have abstained from food or drink, and the going down of the sun is keenly welcomed as the end of one day more of fasting.

The secret of the prosperity of Assouan lies in its granite. It is the granite bed of the river at this point that makes the Reservoir possible; here are the granite quarries from which the Dam was built, and from which every ruler of Egypt who wished to raise a monument for all time has drawn his supplies. Nothing that I have seen in this country brings the past so near as these quarries. Here lies a rough-hewn obelisk, just ready to be rolled away; here an enormous block of stone half hollowed into a bath for an Emperor, or a sarcophagus for an Apis bull, designed by some mighty ruler who 'thought in continents,' and recked little of the lives and labours of thousands provided he gratified his whim. But suddenly death or some other fate intervened, and a feebler or more merciful generation has never taken up the work. You may see the marks of the wedges on some great face of rock, as fresh as if it was only yesterday that Pharaoh's workmen had driven them in and poured water on the wood till it swelled and burst the stone; down below is the fallen piece still waiting for the mason who never came back to it. Perhaps some of the very stones cut by Cheops or Rameses have been smoothed and planed and set in the Great Dam.

Passing along the raised causeway, down which so many great monuments have been rolled slowly to the river, I came through a long stretch of burning desert to a spur on the northern extremity of the granite hills. Here, unexpectedly, I found myself overlooking the lake formed by the filling of the Reservoir. Graceful lines of palm-trees showed where the banks of the river had once been. Philæ was but an insignificant speck on the blue waters, overpowered by the fantastic piles of granite boulders that hem in the valley. In the far distance rose some lofty hills, crowned by dazzling sand that might easily have been mistaken for snow. The Dam builders have been accused of vandalism, but they have created a standing pool in the wilderness of surpassing beauty. The view of the lake was not the only attraction of the spot; at my feet lay a colossal statue of Osiris, destined for some temple, but never moved from the spot where it was hewn.

Ancient Egypt may well look on with scornful wonder at our pride in our achievements. The Great Pyramid at Gizeh contains three times as much solid masonry as the Dam, cut from these same quarries. Every one of those huge blocks had to be dragged to the river, and carried down 600 miles, before it was hoisted into its place. As an achievement of mechanical power the Great Dam cannot compare with the Great Pyramid;

but when at last I climbed a little hill hard by the river, below the Reservoir itself, and saw the whole length of the great stone rampart, stretching right across the valley, the contrast between the world of Pharaoh and our own came strong upon me.

Pharaoh, to whom time and life were nothing, out of the misery of the forced labour of his subjects, raised a perfectly useless monument of his own folly; yet he achieved his object, and made his tomb one of the wonders of the world. We, with the free labour of voluntary workers—paid, fed, and cared for, instead of being driven by the whip—have dared to harness Nile himself. It is a work vital to the interests of millions of dwellers by the river. Yet who can say that the fame of the Pyramid will not endure the longer? Hundreds of years after the time of Cheops a mighty Dam was built in Southern Arabia for a like purpose of irrigation; it lasted for eight centuries or more, and its bursting in 100 A.D. is mentioned in the Koran. Its ruins remain to this day, and show that it was larger than the Nile Dam. Eight hundred years is a long life for a reservoir, but if this one lasts a quarter of that period it will have repaid its cost many hundred times over.

Certainly it looks strong enough to last as long as the Nile itself. Strength, and nothing but strength, shows in every stone of it. Square, solid, and massive, it runs from shore to shore in an absolutely straight line, without the slightest attempt at any trace of ornament or decoration. Clearly, effect has been the last thing thought of.

Even the sluice-gates have absolutely plain rectangular openings, and it detracts from the symmetry of the design that, owing to engineering exigencies, they are not all of one height, but run in sections of five, some of them lower than others. The top of the wall runs in a simple, unbroken level; there is nothing to catch the eye as it travels up the steep face of masonry except the slightest change in the angle of the slope to a nearly complete perpendicular. The wall simply leaves off because its builders thought it high enough for the present. Along the broad surface of the summit runs a tramway, with plenty of room for a man to walk on either side, flanked by perfectly plain, solid parapets as high as the waist, and more than a yard thick, of a piece with the masonry below.

On the eastern side the Dam is unostentatiously built into the living rocks; no arch or pylon marks its start. On the western side it is flanked by a ladder of four immense locks, set in a mountainous embankment. The gallows-like arms of the drawbridge, hideous in appearance, but a marvel of mechanical ingenuity, over the upper gate of the highest lock are the only break in the long, unrelieved level. The hard gray colour of the granite strengthens the general impression, though in time every part exposed to the action of the water will be coated with the shining black varnish which the Nile mud always lays on granite, and it will look exactly the same as the natural bed of the stream.

How different all this is from the prettiness of the Delta Barrage, with its brickwork, originally designed and built under French influence, adorned with archways and towers, of which the lovely garden, with its flowers and shrubs, its green lawns and leafy trees, on the tongue of land which it crosses, seems a natural and appropriate part! The Assouan Dam is the work of a practical, unimaginative race. Its builders have had before them the problem of harnessing the great river with a yoke that cannot be broken; they had to hold up a reservoir containing 100,000,000 tons of water, so that for 140 miles the river is turned back upon itself; and they have succeeded.

In December all but a few of the sluice-gates are shut, for the reservoir has to be filled. But imagine it at the height of the flood, when the collected rainfall of half a continent is crashing past at the rate of nearly 1,000,000 tons a minute, and through each of the 180 openings shoots a solid cube of dark water to dash thundering in clouds of white foam on the rocks below, and rush tumultuously down the swirling slopes of the cataract. Think of the huge bulk of water held up when the reservoir is full. Then you will understand something of the difficulties of the work. Solidity and strength could not but be the first and overpowering idea in the minds of the builders. The longer you look, the more you are impressed. The vast dimensions of the Dam grow upon you from moment to moment. There is, after all, a fierce beauty in those uncompromising features, grimly set in the determination to hold the river in bondage. The massive structure is in harmony with the forces of Nature. Feeble and puny it may be compared with even the least of their handiwork, but it is impossible not to feel that its builders have been inspired with a spark of the same creative power. They drew their plans, and dug and built and strove their best to control the great river. They have succeeded because a portion of that spirit of Nature against which they struggled has passed into their work.

Is it too much to hope that a scheme of decoration may yet be found in consonance with these ideas? Some day the Dam will be raised to the full height of the original design, thus doubling the present capacity of the Reservoir. Then will be the time to finish the work magnificently, and make of it a stately monument, to be the glory of Egypt as well as the foundation of her material prosperity. The subject is worthy of a great artist. Only a scheme conceived on grand lines, perfectly simple and bold, can have the least chance of success. Anything else would be as ridiculous as a proposal to place a statue of ordinary dimensions on the top of the Great Pyramid. The difficulties are great, and so would be the expense, so great, indeed, that it would be far better to avoid any attempt at decoration, unless the results are to be admirable beyond all question. Are there no possible successors to the architects of Karnak?

The ceremony of inauguration was, like the masonry of the Dam itself, sensible and solid, but it was not impressive. Those who arranged its details had forgotten the vastness of the theatre in which it was performed. From the purely spectacular point of view it was a failure. Egypt has no money yet to spend on functions. Perhaps a better stage management might have made a better display without any greater expense. The material benefits of English rule would be appreciated none the less gratefully for a little gilding. But the Englishman in Egypt has had other things to think about than the organization of what the native would call 'fantasias.' So he fell back on the established custom of his own country. Wherever he goes he carries with him law and order and equity and righteousness and common-sense, and he also carries a peculiar kind of public ceremonial routine. Everybody knows it. The invited guests arrive in special trains, and perspire in top-hats and frock-coats for an interminable time, until Royalty arrives full of gracious smiles amid a cheering crowd. The distinguished persons pass into a pen carpeted with red baize, where all the notables are assembled. Somebody makes a perfectly inaudible speech, which receives a gracious and inaudible reply. A button is pressed here, a lever turned there, and several extraordinary things begin to happen in consequence. Then a number of good men and true receive some well-earned decorations, Royalty graciously departs, and everybody presses home as best he can, while the band plays the National Anthem. The whole is accompanied by the clicking of innumerable cameras, and nobody quite realizes the importance of the occasion until he reads about it afterwards. At the Nile Reservoir all this happened in due order with the necessary local variations. The officials wore red tarbushes instead of top-hats, and every sensible person carried a cotton umbrella over his head.

But apart from the ceremonial, the scene was deeply interesting. If the day's routine was insignificant, it was because the overpowering presence of the Dam itself dwarfed every other presence. And the names of those assembled there recalled vividly the thrilling history of Egypt during the last twenty years. There were statesmen and diplomatists, soldiers and engineers, men of business and men of letters, all of whom, in some field or other, had done their part in building up the fallen country. Some, too, were there who, submitting contentedly before the logic of accomplished facts, burying old rivalries and animosities, had come in no unfriendly spirit to witness the realization of much that it had once been their policy to hinder. But if all the rest had been absent, the presence of one man, the Chief, whose wise counsel and guiding hand had been everywhere, would have been sufficient to represent all that these twenty years have meant to Egypt. Well might Lord Cromer and the irrigation engineers review their work with satisfaction. To them the Reservoir means the successful culmination of a great policy long and steadily pursued, nothing less than the establishment of the prosperity of Egypt upon a sure and certain basis : for that is what the regulation of the Nile involves. In Egypt, at any rate, they require no formal monument. Their praise stands clearly writ on the face of every cultivated field throughout the country.

CHAPTER X

BRITISH RULE IN EGYPT

At the inauguration of the Nile Reservoir at Assouan, it was an Egyptian Minister of Public Works who read an Arabic speech congratulating the Khedive on the completion of the great work which is to make his name famous among the rulers of Egypt. Among all the flags that decorated the town and the craft on the river, the most infrequent was that of England. A casual observer, knowing nothing of the country, might easily have overlooked the number of Englishmen wearing the tarbush, that red badge of Egyptian officialdom, and gone away thinking that even the presence of the brother of the King of England marked nothing but a compliment paid by one great Power to another. He might well have been astonished to be told that the Dam, which will confirm and increase the prosperity of Egypt, is no less an evidence of the stability of British rule. It is just possible that immediately after our first occupation we might have been able to evacuate the country—not, indeed, without danger to our hold upon the highroad to India, or without detriment to the true interests of Egypt, but at least without loss of honour to ourselves. Since then, in spite of the efforts of our statesmen at home, it has become more and more impossible.

Among the guests at Assouan there might have been seen a quiet-looking old gentleman, with a gray beard and bushy whiskers, beaming benevolently through gold-rimmed spectacles. His figure was that of a man, once sturdy and square-set, over whose head had passed years of ease and good living. At the first glance, in his frock-coat and tarbush, he looked like any other comfortable Turkish gentleman. Yet there was no one present with a more interesting past than he. For this was Mukhtar Pasha, who bears the proud title of Ghazi, 'the victorious,' the hero of the Caucasus in the Russo-Turkish War. He came to Egypt in 1887 as special Turkish Commissioner, to arrange for the British evacuation under the Drummond-Wolff Convention. It is well known how France and Russia at the last moment intervened to prevent Turkey from ratifying the agreement. A special Providence guards the British Empire against the efforts of its rulers. But Mukhtar remains as Turkish Commissioner in Egypt without duties, and probably without pay, a reminder of past eccentricities of British policy.

We have travelled far from the days when it was seriously proposed by Conservative statesmen to make Turkey responsible for civilization and good government in Egypt. To-day no one in his senses could wish to put an end to British rule. Let a man start from Assouan and survey the great series of irrigation works—the Reservoir, the Assiout Barrage, the Regulator of the Ibrahimiyah Canal, the Koshesha Regulator, the

Barrages and Weirs at the point of the Delta and at Zifta; let him examine the intricate system of canals, siphons, wheels, drains, dykes, and sluices, by which the water is distributed over the cultivated lands, and let him reflect on what would happen if all this were left in Egyptian hands. Inevitably, sooner or later, the whole thing would come to ruin, and the greater the height of prosperity to which the country has attained under the system of perennial irrigation, the greater would be its fall. Egypt has been called the classic land of baksheesh, and it will not lose its character in a generation or two. Imagine a Government in need of money; what better thumbscrew could an Oriental despot wish for than the command of the water-supply? When a land-owner knew that he could be ruined by the shutting of a sluice-gate, he would pay anything without a struggle. The golden goose would be killed in every direction. Even under a well-intentioned Government it would only be regarded as natural for a local official to make free use of such unrivalled opportunities to supplement his pay. Corruption is not a vice in Oriental eyes; it is the habit of centuries. That is why, with every extension of scientific irrigation, the need for European supervision becomes greater, and since we can allow no Power but ourselves to hold Egypt, European means British.

An amusing instance of the native way of looking at such matters occurred in connection with the making of the Reservoir. A certain sum was allotted for the compensation of those who would be flooded out of house and home. Some of the

Assouan people saw their chance. It does not take much to build a native house. In a short time the Government inspectors had the pleasure of being shown a number of brand-new buildings on the foreshore, with the whitewash still wet upon them. Great was the disappointment of these ingenious speculators. In another case a man was building a boat; he was repeatedly warned to move his work, but would not. Only when the water began actually to rise did he become seriously alarmed, and sent down a letter of remonstrance. Again the position was explained to him; but again he advanced his request that the gates might be opened, accompanied with a hint that he might be willing to make some small acknowledgment of his obligation. Realizing at last the futility of his demand, he exclaimed in despair, 'What is the use of your Dam if you cannot let the water out to save my boat?'

The two great principles of British policy in Egypt have been irrigation and low taxation. Irrigation is the vital necessity to Egypt. Other departments of Government, however urgent their claims might be, have had to wait and to be starved until the reproductive works could be established and set going. Every penny that could be spared had to be cast upon the waters. At the same time we found the people overburdened with taxation on account of the Public Debt, and also bound hand and foot in the meshes of the usurer. We could not, however powerful we might be, hope to hold securely a country with 11,000,000 of population, unless we have something else than mere force to look to. To the Egyptians we are aliens by race and religion, we have no ties of custom or intermarriage; we have nothing but their material interests to appeal to. If we can make them prosperous, if they can save money without fear of confiscation, if we can secure to them the fruits of their labour, we have done a great deal to strengthen the basis of our rule.

There is no doubt that the fellaheen do appreciate the benefits of British rule; it would be strange if they did not. The corvée is gone. Not only has taxation been enormously reduced in amount, but its method of collection has been made equitable and regular. Whereas in former days the tax-collector was in league with the money-lender, and contrived to demand his payments at times when it was certain that the cultivator of the soil would have no money, and would have to pledge his growing crops to raise the amount, now the time of payment is adjusted to suit the harvests. Moreover, the new Agricultural Bank lends to the fellaheen at a rate far below that demanded by the Syrian, Greek, or Coptic usurers; the State gives a small guarantee, and the Government tax-collectors collect the interest and instalments of the loans at the same time as the taxes. The Egyptians are naturally a thrifty people; they are taking advantage of this plan, and are very punctual in their repayments. When the scheme was first started, it was met with grave disapproval by the professed economists. The fellaheen, they said, will borrow first from the Bank, and will then execute a second mortgage, and borrow more on the same exorbitant terms from their old bloodsuckers. But Lord Cromer's keen insight into the character of the people has once again been justified.

It is true that a generation is growing up that knew not Ismail. These have lived only in the new order of things, and have no personal reminiscences to sharpen their enjoyment of the present; but even so the number of those actually discontented must be few, and the number of those who would carry their discontent into action still fewer. More than this, perhaps, it would be foolish to expect. In the East a reforming nation could not be really popular, except among fighters. And the Egyptians are no fighters—they are peaceable people, who love their homes; no one joins the army except by the necessity of conscription. Even the reformatory school-boy cannot be induced to volunteer for so much as the band. Honour and glory are nothing to them; they seek no bubble reputation at the cannon's mouth. Again, the Englishman endeavours to establish impartial justice in the Courts. That is all very well if your opponent is a richer man than you; but what if you could have outbid him quite easily? It is a good thing to be free from the fear of being bastinadoed and fined because your neighbour has given false evidence against you, or because he has influence with the police. But if you, which is at least as likely, wish to do the same to him, how do you profit by the reform?

Then, too, the alien rule may be just and righteous and full of solid benefit, but it is dull. In old days a man might be maltreated and flogged, he might have his property confiscated, he might even lose his life, according to the whim of his ruler, but the same whim might equally make him Grand Vizier. Such fluctuations appeal to the

Oriental imagination. A veiled Protectorate must always be something of a mystery to an Egyptian; the personal rule of a despot, Effendina, the Lord of all, is much more suited to his instincts. At any rate, it is difficult to discover much outward manifestation of an appreciation of British rule; but it would be wrong to argue too strongly from that. The fact is rather, indeed, a proof of the lightness of the governing hand. The Egyptians know very well that we shall never resent any opposition; they have confidence in our forbearance. But if time should bring a change once more—and there have been many changes in the past—it would be an evil day for those who had been too open in their support of English rule. It is well to be on the safe side.

If you take an intelligent and prosperous Egyptian, old enough to remember the days of Ismail—for example, a lawyer who has saved enough money to make a considerable investment in land, the prime ambition of every native, a man who speaks a couple of European languages, has had a good education himself, and is very likely sending his son to Oxford or Cambridge—and question him upon British rule in Egypt, he will probably tell you something as follows:—

'No one who has eyes to see can question the benefit of the British occupation. The country has attained such prosperity as never before. We knew very well that no other European nation would have ruled us with such a single eye to the well-being of the natives. We realize the devotion and ability of the British officials. We would rather have you than any other rulers, and we are well aware that if you went we might easily become subject to a King Stork. In such a case your popularity would become enormous. Doubtless we should clamour for your return. But we cannot help dreaming of the glories that might have been ours if our Khedives had not wasted their chances. Who can say how great an African empire might have existed? Long before European nations began to take a hand in the partition of Africa, we held the whole valley of the Nile to Uganda. With a wise Government and such a starting-point, what limits could have been set to our dominions? By the folly of our rulers we squandered it all, and came to ruin. You have drawn us out of the pit, but you thrust your benefits upon us at the point of the bayonet. In spite of them, and perhaps unreasonably, we sigh for rulers of our own faith and race, and we would sacrifice something of our prosperity if we could feel ourselves the authors of what remained.'

Such longings are natural and creditable, and if the men who feel them were capable or numerous enough to form a real governing class, the prospects of the future might be different from what they are. But they are not; and these dreams must remain dreams—for some time to come, at any rate. Such discontents are the inevitable outcome of the progress of our educative work in Egypt. Nothing illustrates it better than the recent movement among the Egyptian officers of some of the Soudanese battalions. These men had been trained on a British model; they had gained much experience in the stern school of actual warfare, and yet they found the higher ranks of the Service barred to them, and filled by a succession of British officers, younger and less experienced than themselves. Their discontent was natural, but their disappointment was also inevitable. In spite of their training and experience, to have given them the promotion they wished for would have been to ruin the efficiency of the army. You cannot make bricks without straw, however scientific your methods.

It would be very easy to exaggerate the importance of such murmurings. They are as nothing in the face of the rising tide of prosperity which has come to the mass of the labouring population as the result of our rule, and which is its overwhelming justification. In the Delta provinces, in Middle and Upper Egypt, in the Fayoum, the undeniable facts rise up and confront you. Wages have increased in some places as much as 50 per cent., and with the rise has gone an enormous improvement in all the conditions of life. The fellaheen are building better houses, they are better fed, disease is less, they are happier every way. And as labourers they well deserve it. Many faults they have, but nowhere in the world can a more industrious, patient, and hard-working people be found.

In the Mosque el Azhar, the Mohammedan University of Cairo, the interpretation of the Koran is the principal subject of study, and it is said that weeks and months, and even years, are spent by professors and pupils in subtle and ingenious dissertations on such a question as, Who is your neighbour? Is a man living over your head more worthy of the name than one who lives next door? And so on. If in these reforming days these pundits turn their attention to politics, they will find an almost equally insoluble problem in attempting to define the exact nature of British rule in Egypt. To them the question may safely be left. But while the learned few are labouring through its intricate maze with the most agreeable lack of success, the unlearned many will have their own simple answer. They only know the thing was done; it matters nothing by what authority. The water came to them regularly in due season, and the wilderness was made to blossom like the rose.

CHAPTER XI

SCHEMES FOR THE FUTURE

The completion of one programme by the construction of the Reservoir and the works dependent upon it does not mean stagnation. Irrigation is an ever-growing and ever-living science. There can be no standing still. Even in the ordinary routine there are thousands of details always pressing forward for fresh consideration. It takes but a few years for a daring innovation to become old-fashioned. The system of perennial irrigation in Egypt is so young that there is still much to be learnt. It means a totally new style of agriculture. Nothing but constant experiment and constant watchfulness will enable the engineers to distribute the water to the best advantage. The best methods of reclaiming new land, and of preventing the deterioration of the old, the maintenance and improvement of existing locks, weirs, regulators, barrages, and other masonry, the care of canals and drains—these and an infinity of other matters fill the life of the irrigation engineer, so that, always interesting, it is also at times even exciting. Decisions of great and immediate importance have to be constantly and promptly taken. Their duty touches very nearly the lives of all who dwell and labour within their districts. There can be no question of any folding of the hands to slumber. But beyond all this, it is the peculiar good fortune of those whose care is the water-supply of Egypt that they have to travel in thought or in fact over half a great continent, and discuss schemes of a magnitude and extent enough to stagger the imagination of the boldest dreamer. The Anglo-Egyptian army at Omdurman opened an entirely new chapter in the history of the control of the Nile.

No one can hold Egypt securely unless he holds also the whole valley of the Nile. The sources of the river in hostile, or even in indifferent, lands must always be a grave cause of danger, or, at the best, anxiety. If tradition be correct, the Abyssinians on more than one occasion made use of their position on the Nile as a powerful lever in negotiations with Egypt. Vansleb, the Dutch seventeenth-century traveller already quoted,* is not perhaps perfectly trustworthy as to his facts, but at least he is evidence as to the tradition. He had in his possession, he says, a copy in Arabic of a letter written by David, King of Ethiopia (*i.e.*, Abyssinia), surnamed Constantine, to 'Abu Seid Barcuk, King of Egypt, in the year of martyrs 1193, in which he threatens in two distinct places to turn aside the river Nilus and hinder it from entering into Egypt,' so as to cause all the inhabitants to perish with hunger, if he 'continues to vex the Copties.' Vansleb also says that 'the King of Ethiopia hindered the current of Nilus and turned it out of Egypt in the days of Mostanser, one of

* *Cf.* p. 71.

the califfes of Egypt, which obliged him to send the Patriarch of the Copties with rich presents to the King of Ethiopia, to entreat him to take away the bank, which he had raised to turn aside the river. The King of Ethiopia having granted this request for the sake of the Patriarch, the river increased in one night 3 cubits, which was sufficient to water the fields.' Doubtless the tradition is founded on fact; if the country in question had been occupied by a Power with any engineering skill, there is little doubt that this method of pressure would have been oftener adopted.

The danger is by no means an imaginary one, and this is recognised in the latest treaty between the Government of the Soudan and Abyssinia, concluded May 15, 1902, in which it is laid down that no work shall be constructed across the Blue Nile, Lake Tsana, or the river Sobat, which shall arrest the flow of their waters into the Nile, except by mutual agreement.

If even now it is thought necessary to make formal treaty arrangements on the subject with Abyssinia, much more real would be the apprehension if the Soudan Government were hostile to Egypt. But quite apart from any hostility, an important question might arise. We have seen what great efforts have been made in Egypt under the system of perennial irrigation to secure a good supply of water in the summer months; it is then that the river is at its lowest; the best of reservoirs can only supplement the natural flow of the river. At the very time that it would be easiest to do it, an interference with that natural flow would be most disastrous. The mere employment of a considerable amount of water on irrigation in the Soudan during the summer months might have a very serious effect. The possession of the Soudan practically guarantees to Egypt the safety of her water-supply. The fact that the sources of the White Nile are in purely British territory, and that in the Soudan itself the British flag flies side by side with the Egyptian, gives thus a more and more permanent aspect to the British position in Egypt.

It is calculated that at the present moment another 3,500,000,000 cubic metres of water over and above the present summer supply, including the Assouan Reservoir, would amply suffice for all possible requirements in Egypt. When the Dam is raised to its full height, about 1,000,000,000 cubic metres more will be provided. If only Egypt were to be considered, the remainder of the water required could be stored in a similar reservoir built somewhere above Wadi Halfa. Egypt could then be developed to its very fullest extent, but 'such a work,' as Sir W. Garstin remarks, 'would leave untouched the countries bordering the river to the south. Their interests must be safeguarded by such a scheme as will insure them a proportional share in the prospective benefits.'

From every point of view the interests of the Soudan are vital to Egypt. Bitterly has she suffered from the neglect of those interests in the past. Misrule and oppression caused the Mahdi's rebellion, with all its burden of shame and suffering. With a more intelligent enemy the loss of the control of the Nile might have meant irretrievable disaster. She knows now that a prosperous and contented Soudan is the best safeguard against rebellion. Upon England, too, rests the responsibility for the welfare of those vast provinces, once so shamefully abandoned to barbarism and ruin. English arms have retrieved that disgrace. It would be impossible to allow the Soudan to get nothing and Egypt everything, even if such a course were not certain to be fatal to both.

At present far-reaching schemes are being carefully inquired into, under the guidance of Sir W. Garstin. Fortunately for Egypt, she possesses in him an adviser of unsurpassed experience, cool, careful, and level-headed, upon whose final judgment she may confidently rely. He has already made a report, after personal observation, upon the White Nile and its main affluents, and he has recently returned from a journey to Lake Victoria Nyanza and Lakes Albert and Albert Edward, the results of which have not yet been published. Another expedition was also sent in 1902-1903 to explore Lake Tsana and the upper course of the Blue Nile. A mass of information is being collected and quietly digested, and in due time the policy to be pursued will be settled on. For the moment there is no occasion for hurrying. It will take time for Egypt to make full use of the advantages of the Assouan Reservoir. The Soudan, though rapidly progressing, is hardly yet in a position to reap the advantage of such changes on a great scale. When the engineers have completed their observations, there will still be the financial aspect of the question to take in hand. Schemes involving the regulation of 2,000 miles of river cannot be subjected to too great scrutiny before arriving at a decision. It would be rash to speculate on what course will be finally adopted, before all the data are known, but it will not be out of place to give some indication of the rival projects in view, and to glance at some of the arguments for and against.

All the schemes aim at establishing a control over the head-waters of the Blue or the White Nile, and so increasing the supply in the river when it is most scanty. Anyone can see that the Blue Nile depends upon Lake Tsana, which is across the Abyssinian frontier, and the White Nile upon the Lakes Victoria and Albert Nyanza. Proposals have been made for regulating the outflow from each of these lakes, or from them all, and also for preventing the waste of water which now takes place in the swamp country south of Fashoda. Up till now the plan of building a dam at the point where the Blue Nile emerges from Lake Tsana has been most in favour.

It has several great advantages. This lonely lake is situated some 6,000 feet above sea-level, in a wild and desolate country in Northern Abyssinia. It is 3,300 square kilometers in extent, and very deep. During six months in the year there are very heavy rains in the area which drains into the lake. Between Lake Tsana and Khartoum, a distance of 1,350 kilometres, the Blue Nile falls over 1,000 metres, and its rocky bed makes it very suitable as a channel for carrying the water. It is deep and narrow, with high, steep banks, and so the less liable to evaporation. For every metre by which the level of the lake was raised, 3,300,000,000 cubic metres would be stored, and it is obvious that, if the level were raised

4 or 5 metres, there would, after making a proper deduction for evaporation, be ample water to supply the greatest demand in Egypt, and also plenty besides available in the Soudan. There are also strong arguments for irrigation works on the Blue Nile. After the Cataracts of Rosaires, which are roughly speaking about halfway between Khartoum and Lake Tsana, the river passes through fertile plains consisting of rich alluvial soil, extending to a great distance on both sides At present these lands yield only small returns, dependent as they are upon a somewhat capricious rainfall. With proper irrigation they would be most productive, and the water coming down from Tsana might be utilized, without any great difficulties, by a system of canals starting from this point, assisted by barrages and weirs. Further, the river would thus be rendered navigable at all seasons of the year, and would thus become an effective trade route in the most promising part of the Soudan. The great objection is that Lake Tsana lies wholly within Abyssinian territory. Nothing could be done without some very definite arrangement with the Emperor Menelik, and one that would be irrevocable by any successor of his who might be less friendlily disposed. Nothing, in fact, would be absolutely satisfactory unless complete sovereignty over the Reservoir district were assured to the Soudan Government. In view of the fact that the whole of this region is perfectly valueless to Abyssinia, or, indeed, to anybody else, except for purposes of water-storage, it does not seem impossible that by some cession of territory, or other compensation elsewhere, a satisfactory arrangement

may be come to. But even so, the remoteness of the lake and the difficulties of transport would prevent the immediate realization of the scheme. Without a railway from some point in the Nile Valley it would be practicably impossible to collect the necessary materials and supplies, and such a railway will not be an affair of this year or next.

Precisely on the points where the Lake Tsana scheme is weakest the schemes for utilizing one or both of the great lakes of Uganda are strongest. Precisely where it is strong they are weakest. Any works damming the exits of Lake Victoria Nyanza or Lake Albert would be wholly within British territory. They would also be in a region much more easily accessible from the sea. Lake Victoria has already railway communication with Mombasa, and the transport of material, and also of coolie labour from India would be comparatively easy. In the case of Lake Albert, a railway would have to be built from Lake Victoria; but this would be of service to the country apart from any irrigation works, and would, indeed, practically establish through communication by boat and railway between Mombasa and Alexandria. But here their advantages over Lake Tsana end. The whole region is very much subject to earthquakes. The strain to which great masonry works would be subjected might be very severe. The country through which the White Nile passes is very unsuited to large irrigation works. Much of it is swamp, and the low slope of the land is ill-adapted for canals. The soil, too, is poor and light compared with the rich alluvial tracts on the Blue

Nile, which would of course receive no benefit. Practically only the provinces north of Khartoum would receive any benefit from the increased supply, and in them the cultivable land can never be more than a mere strip along the river, so circumscribed are they by the desert ridges. In any case they would be equally benefited by water coming from the Blue Nile. Besides this, the negro population living on the White Nile is very much less advanced, and less likely to form an industrious agricultural population than the inhabitants of the Eastern Soudan.

Nor is the bed of the White Nile well adapted for carrying the water. From Lake Albert to Khartoum the distance is 2,100 kilometres, and the total drop is only 300 metres. It flows slower through a hotter country. Except between Duffile and Rejaf it is very wide and shallow, without any banks to speak of, and with a sandy or muddy bottom. For at least half its course it runs through swamps, and between Bor and Lake No alone it is calculated to lose half its volume by dissipation and evaporation in the marshes. But for the important contribution made by the Sobat, the volume reaching Khartoum would be very much smaller than it is. If, then, a considerable extra supply of water is to be brought down from the equatorial lakes, the scheme must involve a permanent improvement in the channel of the Bahr el Gebel, as the upper portion of the White Nile is called.

This, according to Sir W. Garstin, can only be effected in one way, by embanking the river for its whole length between Bor and Lake No, a distance of 624 kilometres. When it is considered that all supplies would have to be brought from Khartoum, 1,000 kilometres distant; that during four months of the year work is impossible, owing to the incessant rains; that the local tribes can never be relied upon for labour; that the climate is exceedingly bad and unhealthy at all seasons; and that the actual engineering difficulties in making the banks would be by no means small, some idea may be formed of the cost of such an undertaking. It is estimated that to complete it in five years' time would involve an expenditure of £3,700,000. Another proposal, which is really independent of the Reservoir question, is to use the Bahr el Zeraf to carry the extra summer supply at present wasted in the marshes. The Bahr el Zeraf is a branch of the Nile, taking off near Shambe, and entering the river again below Lake No before the junction of the Sobat. The cost of preparing the Bahr el Zeraf channel by means of dredging and embanking is estimated at £1,250,000, and in addition certain supplementary works would be required.

Until the result of Sir W. Garstin's observations on the Albert Lake are known, it will be impossible to make any accurate comparison of the qualifications of the two lakes Victoria and Albert for being utilized as reservoirs. Lake Victoria, which lies between the two 'Rift Valleys,' is encircled by a low and shelving shore. It covers approximately about 70,000 square kilometres. The Somerset Nile flows out of it, and finds its way into Lake Albert. Lake Albert is the northernmost of the chain of lakes in the western Rift Valley, and into it drains the Albert Edward Lake by means of the Semliki River, partly fed by the glaciers of Mount Ruwenzori. The Nile issues from it at the northern end, not so very far from where the Somerset Nile enters. Lake Albert is surrounded by mountains and cliffs. There is comparatively little flat shore round it. Its area is roughly about 5,000 square kilometres. For every metre that the surface of each lake was raised, Lake Victoria would store 70,000,000,000 cubic metres, and Lake Albert 5,000,000,000. These enormous figures are enough to show that, even when the largest allowance has been made for loss by evaporation, dams of no great height on either lake would suffice to store quite sufficient water for all possible needs in Egypt, and all of the Soudan that could be touched by water coming down the White Nile.

A very small rise in the level of Lake Victoria would give a very large reserve of water, and therefore the works regulating its outflow would be of less dimensions, and so, presumably, less subject to damage by earthquakes. Being nearer the sea, they would, moreover, be easier of construction. But Sir W. Garstin so far totally rejects the idea of Lake Victoria. He says:

'The amount of water which could be stored even by a very small rise in the levels would be far beyond any possible requirements. This lake may consequently be omitted altogether from any projects for water-storage. Much of the country adjoining Lake Victoria Nyanza is densely populated, and the villages are situated close to the shores. Any considerable rise in the water-levels would flood a large and populous area of country. It must not, moreover, be forgotten that about half the area covered by this lake falls within German territory. As the inhabitants of the southern half of this lake would derive no benefit from such a reservoir, it is quite conceivable that they might view any such proposal with disfavour.'

He proceeds to argue that Lake Albert, on the other hand, is specially adapted by its conformation for the purpose.

'With a regulating dam at a point on the river below its exit, the Albert Lake could well be used to store up water during the rainy season, which would be discharged into the river during the months of low supply. In this way a double purpose would be served: the volume of the river in flood would be diminished, and in summer would be largely increased. The lake has an enormous catchment area, and it seems probable that its levels could be, without serious difficulty, raised to the required height.'

In all probability the arguments will be found to be conclusive in favour of a dam at the exit to Lake Albert, but the reasoning against a regulator for Lake Victoria does not appear at first sight very cogent. Sir W. Garstin tells us that three years ago the mean water level of the lake averaged some 8 feet lower than it did twenty years earlier. But between January 1 and June 1, 1901, the level rose 3 feet 3 inches. When such fluctuations already occur in the ordinary course of Nature, it would seem that some very useful regulation might be carried out without causing the least inconvenience to any dwellers on the shore, German or otherwise. All these remarks, however, were written quite tentatively by Sir W. Garstin before he had personally examined these regions, and his next report will be certain to give fuller information on the matter. Perhaps in the end both lakes will be subjected to the yoke of the engineers.

Such in brief outline are the gigantic schemes which are now engaging the attention of the irrigation authorities of Egypt. Great as are the achievements of the past, they look almost petty before these visions of the future.

CHAPTER XII

THE SUDD

If the White Nile carried its waters in a channel in any way resembling that of the Blue Nile, there would be small cause for anxiety over the summer supply of Egypt. As far as Lado, indeed, the Bahr el Gebel, or Mountain River, as it is called, does flow with as good a slope and as sound a rocky bed as any Egyptian could desire. Then it enters upon the great plain which extends practically as far as Khartoum, but until Bor is reached it has not to encounter more than the ordinary troubles of a flat country. The swamps are still well above the summer level of the river, and the loss of volume is not great. At Bor, however, the real difficulties begin, and the 'shorn and parcelled river strains' miserably along for some 350 miles through the real sudd country. The river that leaves Lado in April or March with a volume of 600 cubic metres per second is reduced to less than half this amount by the time it emerges from the maze on the further side, and that, too, although it has nominally received at Lake No through the Bahr el Ghazal the drainage of a vast province.

Lake No is the remains of an inland sea which

once covered all this desolate region. Gradually this lake became silted up with peaty deposits, and formed a series of swamps and lagoons through which the Bahr el Gebel from the south and the Bahr el Ghazal from the west wander in a series of loops and curves without any certain banks. The smallest rise in the level of the river floods an immense tract of country, and at all times of the year the sudd region is one gigantic evaporating pan. The Bahr el Ghazal, so far as any active contribution to the volume of the Nile is concerned, is practically lost altogether, though by spreading out over the marshes it fills a space which the Bahr el Gebel would otherwise occupy, and so helps to form a sort of reservoir, which prevents any rapid fall in the level of the White Nile.

All these swamps are covered with a dense growth of reeds and water-weeds. Of these, the most important is the papyrus, which grows in great abundance, and often reaches a height of 10 to 16 feet. Once the papyrus was common in Egypt, but now it is not found north of Abu Zeid, 190 miles south of Khartoum. In Europe it is not found, except on the river Anapo in Sicily, where, however, it does not attain anything like the same luxuriant growth. There are also great stretches of the reed called in Arabic 'um soof,' or the mother of wool. It is no wonder that the party sent by Nero to explore the river was intimidated by these interminable forests of papyrus and reeds, and turned back. No travellers have a good word to say for this country, unless they only pass through it extremely quickly. It is monotonous and desolate in the extreme, the air is always hot and steamy, and after dark mosquitoes rage in countless myriads.

As far as Shambe the banks of the river are fairly well defined, but there are numbers of breaches in them, through which the water spills into the marshes. In a distance of seventy-five miles Sir W. Garstin counted 129 of these breaches, usually about 4 yards wide; nearly all of these were on the eastern side. These channels are deeply cut with vertical sides, as if dug by hand, and the loss of water through them is very great. The marshes on the eastern side drain off into the Bahr el Zeraf, which takes off near Shambe, and, cutting off the corner, rejoins the Nile below Lake No. Greatly diminished by this loss, the Bahr el Gebel continues on its way, its bank being often nothing more than a line of papyrus separating it from large lagoons or dense thickets of reeds. As long as the channel is clear, tortuous and winding though it is, a great deal of water is able to pass; but under certain circumstances the channel becomes blocked by bars of solid vegetable matter, called sudd. Upstream of the obstruction the level rises, and, spreading out over the marshes, more and more water is lost in evaporation.

Sudd occurs both in the Bahr el Ghazal and the Bahr el Gebel, but it is of different kinds. The Ghazal sudd is very much less substantial than that of the Gebel. It is chiefly composed of masses of the smaller swimming plants, which grow in Lake Ambadi and other shallow lagoons on that river and its affluents. But on the Gebel the principal constituents of the sudd are the papyrus and the 'um soof' reed. Both these plants grow with their roots bedded in the soil below the shallow water of the lagoons. Strong gales are prevalent in these latitudes, and these loosen the roots, so that, if the storms are accompanied by flood, large masses of the plants are set free, and begin drifting over the lagoons. Quantities of earth remain clinging to their drifting roots, and thus they are driven backwards and forwards by the wind. The storms come generally just before the flood, and thus there is plenty of water for the reeds to float in. Eventually they are driven, whole acres at a time, into the channel of the river, and are carried down the stream. But the channel being very narrow and winding, they soon get caught up, and a block occurs. New masses come down, and these are sucked by the stream under the original block, till a dense mass is formed, sometimes as much as over 20 feet thick, and so solid that a man can walk over it. If left to themselves, these blocks may last a very long time, and cause the river to change its course; or else by increased pressure the block is burst, and a great wave passes down the river, which sweeps all other blocks away. Years of high flood are generally followed by a great deal of sudd.

It appears that the conditions on the river between Lake No and Shambe are much more favourable to blocking by sudd than they used to be. In 1840 Mehemet Ali sent a scientific expedition under D'Arnaud to explore the White Nile. At that time the main stream through the sudd country was much bigger and stronger. No mention is made of the Bahrel Zeraf at all, which looks as though it was of much less importance, at any rate. They noticed, however, a number of openings in the banks between the river and the swamps, some natural, but some artificial, made by the natives fór fishing purposes. It is plausibly argued by Sir W. Willcocks that these openings were later widened by the natives to enable them to escape in their canoes from the slave-dealers, and then by the slave-dealers themselves, escaping in their turn from Government patrol boats. All these spills would drain into the Bahrel Zeraf, which gradually became a navigable stream in its lower part. In 1863 Sir Samuel Baker ascended the Bahr el Gebel without any difficulty ; but in that year there was a very high flood and a great deal of sudd. Returning in 1865, Baker found the river north of Lake No considerably blocked. In 1870 Baker was unable to ascend the Bahr el Gebel with his expedition to Equatoria, and went by the Bahrel Zeraf. He cut through the sudd in that stream with prodigious labour, and had finally to cut a ditch through 2,000 feet of solid clay Even then the Bahr Zeraf level was so much below that of the Bahr el Gebel that he could not get his boats through until he had dammed the Zeraf behind them so as to lift them up. In 1873 he says he found the canals much improved by the force of the stream. All this naturally contributed to weaken the power of the main stream and render it very liable to sudding.

In 1874, when the river was low, the sudd was cleared away by Ismail Pasha Ayoub, then Governor-General of the Soudan, and all through Gordon's time in Equatoria there was free passage, but the channel had dwindled very much.

The heavy flood of 1878 brought the sudd down again worse than ever, where it remained until cleared by Marno in 1880. It was in that year that Gessi, on his way back from his province, was caught and suffered such hardships in the sudd on the Bahr el Ghazal. He was only rescued by the appearance of Marno. Unfortunately, no record has been left of the methods which Marno employed in the work, but he spoke of it as quite an easy matter.

Little is known of the state of the sudd during the years of the Mahdi and the Khalifa ; but in 1898 Lord Kitchener found the Bahr el Gebel blocked by sudd immediately above Lake No ; and about the same time Colonel Martyr, coming from Uganda, was unable to penetrate more than 20 miles north of Shambe for the same reason. As soon as the Khalifa had been finally crushed, operations to clear the channel were immediately undertaken. In the first three months a length of 80 miles of river had been cleared, including 5 miles of actual sudd in eleven blocks, besides three more blocks which broke away of themselves. The blocks that came away of themselves represented a very large amount of sudd. After one of them had burst, the floating weed took thirty-six hours to pass a given point. The effect produced by the sudd in damming up the river is illustrated by the fact that when the third block, which was nearly 20 feet thick, was removed, the upstream level fell 5 feet in four days, and all the swamps and lagoons began to drain into the river.

The amount of labour was prodigious. At one place, owing to the enormous masses of um soof and papyrus that kept pressing into the river, no less than eleven clearances had to be made in one year. Five gunboats and 800 dervish prisoners, besides officers and guards, were employed. Sir W. Garstin gives an interesting description of the methods by which the sudd was removed :

'In the first place, the papyrus and reeds on its surface were burnt (curious to relate, these green reeds burn readily when the marsh is dry), and in a few hours' time the surface of the sudd was an expanse of blackened stalks and ashes. As soon as the fire had died down, gangs of men were landed from the steamer and employed in cutting trenches on the surface of the sudd. These trenches averaged from 0·60 to 0·80 metre in width, and from 1 to 1·25 metres in depth—*i.e.*, as deep as the men could work. The surface was thus divided into a number of rectangular blocks some 3 metres by 4 metres. To these blocks were attached, one by one, steel hawsers and chains. This done, the steamer downstream went full speed astern. It invariably took several pulls to detach a block, and in some instances as many as nine were necessary. Both hawsers and chains kept constantly breaking, although the former were calculated to stand a strain of 35 tons per square inch, and the latter as much as 60 tons. As soon as a block was detached from the mass, it was allowed to float downstream. It was curious to see the green reeds and papyrus which had been confined beneath it reappear on the surface of the water. A horrible stench prevailed from the rotting vegetation.'

In 1900-1901 four more blocks were removed. An experiment was made to break up one of the blocks by means of explosives, but it was found that the sudd, though very solid, was too elastic, and the only effect was to make large holes in it and no more. The old method had to be once more resorted to. By the end of 1901 only 23 miles of the channel remained uncleared, but this proved to be the most difficult of all. One block was actually 7 miles in length. Portions of the sudd had rotted, and, sinking to the bottom, completely filled up the channel, so that there was no stream at all, and it was impossible to tow pieces away and float them downstream by the usual process. To get up the river it is still necessary to go round by a series of lagoons, in which navigation is difficult, and operations have been suspended for the present.

The sudd in the Bahr el Ghazal was much easier to remove, and great progress has been made, so that now the whole of that river is free, and boats can even ascend the Jur River to Wau.

It is very difficult to ascertain the precise effect of the removal of the sudd on the water-level in Egypt. But it is certain that as each block was removed a quantity of water drained off the marshes and helped to diminish the fall of the river in summers of very small supply. Most of this water, spread out over a large surface, must otherwise have been evaporated and lost altogether. There can be no doubt that the improvement of the channel has been a permanent gain to Egypt of a very substantial kind, to say nothing of the advantage to the Soudan from the opening up of such an important line of communication. Those who laboured at the task deserve the highest praise. The river was so low that all communication with Khartoum was cut off for several months, and transport of supplies was always difficult. Besides this, the climate is always unhealthy, and the mosquitoes at night are almost beyond endurance.

Under present conditions it will require constant care and watchfulness, especially in years of high flood, to prevent the sudd from frequently obstructing the river. But if the danger is to be permanently removed, the main channel must be so much improved by widening and deepening it that it will carry a greater volume. This will at the same time prevent the water from being dissipated in the marshes, and diminish the chances of any obstruction. The necessity for some work of this kind, if there is to be a reservoir at Lake Albert or Lake Victoria, has already been referred to. The clearing of the sudd is only the essential preliminary to the greater scheme.

One point remains to be noticed. Under the old basin system in Egypt there could hardly be too high a flood, nor did a low summer Nile create any extraordinary difficulties when there was so little summer cultivation. But with perennial irrigation a high flood becomes a matter of supreme anxiety, and the preservation of its dykes is the anxious care of every village. On the other hand, all the schemes for reservoirs aim at increasing the summer supply. If the Bahr el Gebel was trained so as to bring down more water in summer, it would also bring down more in flood, though the swamps would still act as an escape for the waters beyond a certain rise. It might therefore become a matter of great importance to Egypt to actually diminish the supply during the flood, especially as year by year more of the basin land is converted to perennial irrigation. Regulators at Lakes Victoria and Albert would serve in some degree for flood protection as well as for storage against summer use. It is calculated that the complete closing of the outlet of Lake Victoria at the Ripon Falls would only raise the surface of the lake 20 inches in a year. But it would be far more effective if some of the flood-waters of the rivers—*e.g.*, the Blue Nile and the Atbara—which are fed by the rains of Abyssinia could be intercepted before they reached Egypt. A large irrigation during the flood and early winter along the Blue Nile might indeed actually lengthen out the flood in Egypt, whilst depriving it of danger through excess at any one period.

CHAPTER XIII

THE UNITY OF NILELAND

It is no longer possible to think of our occupation of Egypt as merely a stepping-stone on the road to India—'the Englishman reaching far over to his loved India.' Still less can it be looked upon solely as a means for the regeneration of Egypt and the education of her people till they are able to pass from a state of tutelage and stand securely by themselves. Certainly, great strides have been made in this direction. If Egypt were the whole matter we might hope that in time, but not in one or two generations, a race of native administrators might rise up, to whom the affairs of that country might safely be left. But Egypt is only a portion of the great country of the Nile. Looking southward from Alexandria or Suez, the horizon is only bounded by the sources of the Nile, and these do not well up at Assouan, as Herodotus was told. We might hope that the Egyptians would be capable of managing Egypt; but not the most sanguine enthusiast could imagine a period of time sufficient to make them capable of managing the Soudan. A cry of 'The Soudan for the Soudanese' would hardly be more ridiculous. The story of the binding of the Nile, incomplete as that story is at present, makes one thing, at least, perfectly clear, and that is, that all Nileland is one country. No divided sovereignty is possible ; there must be one firm hand over all.

It would have needed a preternaturally keen eye to perceive that from the moment we began to patch the old Barrage the occupation of the whole Valley of the Nile was inevitable. Looking back, it is easy to see how it all followed in logical sequence. Everything depended on the Nile. The more Egypt was developed, the greater grew the need for the regulation of the water. The rulers of Egypt need have troubled little about the fate of countries divided from them by so many leagues of rainless desert, but for the link of the all-important river.

It sounds a far cry from the snows of Ruwenzori, the lakes and swamps of Equatorial Africa, or the rain-swept hills of Abyssinia, to the cotton-mills of Lancashire. The Egyptian peasant, lifting water on to the fields of the Delta, knows that the connection is close enough. We have our own direct commercial interest in holding the Valley of the Nile, and Egypt is still on the road to India. Apart, therefore, from the duties which rest upon us as a civilized Power, we are doubly responsible for the welfare of the people of the Nileland.

Fifty years ago a distinguished English sailor travelled to Khartoum and El Obeid, and published an account of his journey in a little volume entitled 'A Ride across the Nubian Desert.' In eloquent words he describes the wonders of the Nile's course, and continues :

> 'Surely the hand of the Almighty has traced it across the desert that it might be the union of distant nations. . . .

Its mission is not yet accomplished; it is waiting to be the road to civilize Africa. But it is not an Eastern nation, and not the Mohammedan religion that can do it; and I am one of those who hope and believe that Providence will destine it for England. An English Government and a handful of Englishmen could do it. Cities would rise up at Assouan and Khartoum, whose influence would be felt over the whole interior. . . . I know, alas! the spirit of the age is against such thoughts, and there are even men who would wish to abandon our Empire; but I speak the voice of thousands of Englishmen who, like myself, have served their country abroad, and who do not love her least, who will never consent to relinquish an Empire that has been won by the sword, and who think the best way to preserve it is often by judicious extension.'

On many a stricken field the author of these prophetic words, Captain Sir William Peel, V.C., the hero of the Naval Brigade in the Crimea and the Mutiny, proved the sincerity of that love for his country of which he spoke so warmly. And now, after so many years and so many vicissitudes of fortune, an English Government and a handful of Englishmen are grappling with the work on which his heart was set.

PART II

THE NEW SOUDAN

CHAPTER XIV

THE PAST

As with modern Egypt, so with the modern Soudan: the name of the Albanian tobacco-seller is writ large upon the pages of her history. In spite of the ancient connection between Egypt and Ethiopia, in spite of the dependence of Egypt upon the Soudan for her water, the warlike tribes of the south remained for more than 1,000 years free from any attempt at domination by their neighbours. It was Mehemet Ali who, in the year 1819, laid the foundations of the empire which reached its furthest limits under the Khedive Ismail, an empire of which the brief but disastrous history brought nothing but misery and ruin to the Soudanese, and made the Soudan a name of fear and trembling to every Egyptian peasant.

About A.D. 700 Arabs of the tribe of Beni Omr, driven out of Arabia, crossed the Red Sea, and began to settle about Sennar, on the Blue Nile. By degrees these fugitives, reinforced by other tribes, some from Arabia direct, some, it is said, by way of Egypt and the countries further west, swelled to an invading host and permeated the whole of the Northern Soudan, and the original inhabitants were largely converted to Islam. In the Sennar region the two principal negro tribes were the Fung and the Hameg. The conquerors, while they imposed their language and religion on the conquered, seemed to have been absorbed into their ranks; for the distinction between Arab and negro diminished, and the old tribal names reappeared. In 1493 Amara Dunkas, a sheikh of one of the Fung tribes, was recognised as king of all the Fungs, and conquered all the country on both sides of the Blue Nile from Khartoum to Fazokhl. The negroes who remained in the country were merged in the Fungs. The remainder emigrated to the mountains of Southern Kordofan, where, under the name of Nubas, they have ever since maintained themselves in a somewhat precarious independence against the raids of the Arabs of the plains. About the same time was founded the Sultanate of Darfur, which in time extended its dominion to the banks of the Nile. South of these two powers the Shilluks and other negro tribes continued, as before, generally engaged in some petty warfare, and regarded as a convenient reservoir of slaves by their Arab neighbours.

The Fung dynasty lasted 300 years, and attained a very considerable position. About A.D. 1600 in the reign of Adlan, Sennar even became famous for learning, and was the resort of many scientists and philosophers from Cairo and Bagdad. In the reign of King Baadi, 1719-1758, the fame of the kingdom of Sennar reached its height. A quarrel arose with Abyssinia on account of some presents from the King of France which had been intercepted by Baadi. The Abyssinian King invaded Sennar with a great host, but met with the most signal defeat. So great an event was heralded throughout the Mohammedan East. The bazaars of Constantinople and Delhi were alike filled with the renown of Sennar. Once more crowds of learned and celebrated men flocked into the country from every quarter. But it was the last flicker of sunlight before the night fell. Baadi himself was deposed and exiled on account of his bad administration. The Hameg tribe, long subject to the Fungs, began to lift their heads. By 1790 the kingdom of the Fungs had disappeared, and for thirty years the Hameg continued supreme. Fire and sword was their sole notion of supremacy; it was no mere theory with them. The country was utterly given over to anarchy when Mehemet Ali determined to interfere.

The motives that prompted him were many; possibly, ardent irrigationist that he was, the desire to secure the upper waters of the Nile may have been among them; but beyond a doubt, like Cambyses of old, his principal object was gold. Extraordinary rumours were current as to an El Dorado in the south. Every officer and man who took part in the expedition entertained the most extravagant notions of gold-strewn districts. By means of this, and also by securing the profitable traffic in slaves, Mehemet hoped to be able to win sufficient resources to carry out his ambitious schemes in Asia and Europe. It was also very convenient at the moment to find employment for his irregular troops. It would be a mistake, however, to explain his action wholly by such reasons as these. All through his life ran the dream of posing as a second Napoleon. It is curious to find that along with the army he sent a number of learned men and skilled artisans, while it was announced that the object of the invasion was to introduce the benefits of a regular government of civilization. Mehemet Ali had closely studied the methods of his great model.

But of all these designs only one, and that the worst, was destined to be fulfilled. The slave-trade had always flourished in the Soudan. The Arab States were founded upon slavery. Along the great Arbain road, the desert 'track of the forty days' from Darfur to Assiout, yearly caravans containing slaves as well as other merchandise passed into Egypt; the Nile route served the same purpose, and from the Red Sea ports there was a constant export trade to Arabia and Turkey. All this was nothing compared to the dimensions which the trade assumed under Egyptian rule. It became practically the sole trade of the country; it reached like a pestilential blight even as far as the Equator. Bitterly did the Soudan suffer for the Napoleonic ideas of Mehemet, and bitter, but deserved, was the penalty which Egypt had to pay for her misgovernment in the end. Even to-day among the remotest tribes the name of 'Turk' stands for loathing and terror.

The army, commanded by Ismail, son of the Viceroy, reached Sennar without opposition. Thence, accompanied by his brother Ibrahim, he advanced to Fazokhl in search of the famous goldmines in the Beni Shangul. But the gold proved disappointing. Ibrahim returned to Egypt, destined to win fame as the butcher of the Peloponnese in the Greek War of Independence. The

Arabs rose against Ismail, and he had to return to defeat them. Unsuccessful in his quest for gold, he had devoted all his energy to the slave-trade, but his cruelties and barbarities were too much even for a people familiar with the Hameg. The local chief at Shendi, appropriately named El Nemr (the Tiger), invited him to a banquet. While he and his followers, contrary to the precepts of their own Koran, drank freely of the forbidden wine, straw was silently piled high about the house and fired. To a man they perished in the flames.

Meantime, Achmet Bey, the Defterdar, had conquered the province of Kordofan from the Sultan of Darfur. Hearing of Ismail's fate, he marched towards Shendi to avenge him. The story of what happened, as told by the Egyptians, wears an ugly look. The details are wanting, but, at any rate, few were left to tell the other side of the story. At Metemmeh, on the bank of the Nile opposite Shendi, the people sent messengers to sue for pardon. It was granted. But when the Defterdar marched into the town a lance was thrown at him. The pardon was at once rescinded, and a general massacre ensued. El Nemr himself, however, had fled to Abyssinia. He, at any rate, had no faith in Egyptian promises. The Defterdar then marched towards Khartoum, and at Tuti Island another great slaughter took place. It was a bad beginning. To the native Soudanese the distinction between the benefits of a regular government of civilization and the fire and sword of the Hameg must have seemed slight indeed.

To build aright on such foundations would have

been difficult for a nation of born administrators. For the Egyptians it was impossible. There were among the Egyptian Governors of the Soudan honest and righteous men, but, amid a crowd of officials bred and trained in an atmosphere of corruption and slavery, their spasmodic efforts after good only gave fresh opportunities for evil. Military stations were established in various parts of the country for the sake of security, and they became fountains of slave recruits to swell the ranks of the army. The navigation of the White Nile was declared free, and it became the favourite route of the slave-traders. Khartoum, from a village of skins and reeds, rose to be a city of bricks and the capital of the Soudan, but also a convenient and central market for a huge slave-trade. The Abyssinians, who espoused the cause of the Sennar rebels, were beaten back into their mountains, but the savage methods of warfare only brutalized and demoralized the victors. It speaks volumes for the barbarous character of the times that when Adlan, the leader of the Abyssinians, was captured by Kurshid, reputed the best of the Governors of the Soudan, he was immediately impaled. The annals of these years are filled with stories of famine and rebellion, and, to add to the general desolation, cholera and other diseases constantly ravaged the country.

But worse was to come. Down to 1853 the southernmost Egyptian station was only 120 miles south of Khartoum. The annexed provinces of Kordofan, Sennar, and Kassala (or Taka) groaned under oppression and tyranny, but the negro inhabitants of the Upper White Nile and the Bahr el Ghazal were still comparatively unmolested. In that year the English Consul in the Soudan started a trading expedition up the river. Other traders followed, who established stations far up the country. Peaceful trading soon succumbed to the temptations of the slave-trade. For the sake of protection, it began to be found necessary to employ bands of armed Arabs or Nubians. Gradually the European traders disappeared; by 1860 the last of them had sold their stations to their Arab agents. No greater curse was ever let loose upon a country than these human locusts, and the Egyptian Government was directly responsible. Under the shallow pretence of legitimate commerce, trading monopolies were leased to these traders in various districts. The fact that the Government had not a shadow of claim even by right of conquest to the territories leased was no obstacle. All the country south of Kordofan and Darfur along the White Nile or the Bahr el Ghazal was regarded as peculiarly suitable for these nefarious bargains. The Khartoumers, as they were called, because they had their headquarters in the capital, established themselves everywhere, and became practically independent potentates. With their armed bands of brigands they raided the native tribes, and even used them to fight against each other. Only the Dinkas, protected by the impenetrable marshes of the sudd region, and the powerful and warlike Azande, or Niam-Niam, in the south, were able to maintain themselves. But the Bongo, a numerous and peaceful agricultural people, were easily reduced. The fact that they had attained a higher civilization than their neighbours (for they smelted the iron found in their swamps with furnaces of clay and rude bellows, and worked it with stone hammers on anvils of granite, made pottery, and even had some acquaintance with surgery) only made them the more valuable in the slave-market. The Jur, Dembo, and Golo tribes were likewise among the principal sufferers. Anarchy is but a mild term for the condition of affairs.

Some of these unhappy victims may have heard the news on their way to the northern slave-market that slavery was abolished in the Soudan. If so, it cannot have done much to sweeten their bondage or to heal the strokes of the lash, yet a viceregal decree to that effect had been solemnly promulgated at Berber in 1857. It was characteristic of many of the descendants of Mehemet Ali that they inherited great part of his intellectual vigour and wide sweep of imagination without any of his executive capacity. Said Pasha, who became Viceroy in 1854, visited the Soudan and clearly recognised the failure of his foreign empire. A large force had to be maintained to screw exorbitant demands out of a discontented population. Agriculture was depressed, and every other industry was perishing under a system of taxation vicious in itself, and collected by methods which might have made a Verres blush. He determined to evacuate the country. But the burden of the Soudan, so lightly taken up in greed of gain, was not to be so lightly laid aside. Egypt was now holding a wolf by the ears. The officials who battened on the ruin of the country opposed a strenuous resistance. Things had come to such a pass that the sheikhs and notables of the provinces themselves feared that the withdrawal of one devil would be followed by the entry of many. Said contented himself with reorganizing the government and announcing a number of reforms of a drastic and far-reaching nature, and then returned to Egypt. It was the first of many reorganizations and many reforms, all of which were as effective as the decree for the abolition of slavery. The plan of much talking and little doing became a fixed principle of Soudan policy.

The story of the tax on sakiehs (water-wheels) affords a good illustration of Egyptian methods. One of Said's reforms was to fix this tax at 200 piastres per annum. In less than nine years it had risen to 500 piastres. Jaaffar Pasha, who finally raised it to this extent, declared openly that he fixed it at that rate in order to see how much the peasants would really pay, and he hoped after a three years' trial to be able to arrive at a just assessment. It was not a very scientific plan of taxation in any case, but, unfortunately, Jaaffar was removed before the scheme had time to work out, and his successors, absolutely indifferent to his motives, retained the tax, and even further increased it. It was calculated that on average land the tax often far exceeded the net returns for one sakieh. Even a ruined wheel was liable to the full amount, and if an owner returned to it after an interval he was saddled with the whole of the arrears. On the same principle taxes continued to be charged on land and trees that had long since been carried away by the floods. The natural results followed. Many cultivators were ruined and reduced to beggary, others fled the country, and much land went out of cultivation. In 1881 more than 2,000 sakiehs were lying derelict in Berber and Dongola.

By the time Ismail Pasha came to the throne in 1863 it had become abundantly clear that the Egyptians were unfit to govern the Soudan. It looked as though in a few years the whole country would have become a wilderness, totally uninhabited save for a few wanderers, whose sole occupation would be selling each other into slavery. And yet the next few years witnessed an enormous extension of the Egyptian Empire, and Ismail himself enjoyed, until the bubble burst, a great reputation as a genuine and whole-hearted reformer. Nor was that reputation wholly undeserved. Strange compound that he was of vast ambitions but changeable resolution, of far-reaching sagacity but reckless carelessness, a Westerner in the conception of his ideals but an Oriental in every sense in his pursuit of them, he proved himself in his treatment of the Soudan, as in other spheres, to possess many of the elements of greatness. If he failed, it was partly because the evil was beyond cure: the impending catastrophe was too great to be averted. His employment of Baker and Gordon and other Europeans showed that he realized the incapacity of Egypt to perform the task by herself. That was in itself a great step forward; undoubtedly it staved off for a little the day of retribution. His eager support of the project of a railway to Khartoum, first mooted by his predecessor, Said, showed a sound appreciation of the position, though his ineffective attempts to carry it out showed his weakness as clearly. But the whole hierarchy of Egyptian officialdom was rotten to the core. The best of rulers without good ministers is predestined to failure.

To add to a falling house must always be a desperate remedy. No other course seemed open to Ismail, if he was really to cope with the slave-trade. So long as the basin of the Upper Nile remained in the hands of the 'Khartoumers,' the sources of the traffic flowed as briskly as ever, and at the same time the Red Sea ports afforded every facility for export. Accordingly, in 1866 Ismail purchased the districts of Suakin and Massowah from the Turks by an increase of tribute. In 1869 he took a still more important step, and determined to annex the whole basin of the Nile. He invested Sir S. Baker with absolute and despotic powers over the whole country south of Gondokoro. No better choice could have been made. An administrator of the best type, energetic and high-minded, Baker was also no stranger to the scene of his mission. He had already in 1861 conducted an expedition up the White Nile to join hands with Speke and Grant in their investigations of the sources of the Nile, and in 1864 he had discovered the Albert Nyanza.

A strong man was needed. The Khedive seemed in earnest, but he was occupied with the Suez Canal and other matters nearer home. His representatives in Khartoum took quite another view; it was the custom of the Soudan Government to take away with one hand what it gave with the other. Baker's appointment bore the ominous date of April 1, and the fact may well have recurred to him when, on arriving at

Khartoum towards the close of the year, he discovered that the territory he was sent to annex had already been leased by the Governor-General to a couple of notorious slave-dealers. Every conceivable obstacle was put in his way by the officials. But, in spite of all opposition, he organized his expedition, and after a journey of incredible difficulty and labour, for the real channel of the river was blocked by sudd, he reached Gondokoro in May, 1871, and formally annexed it as 'Ismailia.' Next year he passed south, and proclaimed Unyoro an Egyptian province, organized a number of military posts, and entered into friendly relations with M'tesa, King of Uganda. For the time the slave-traffic in these new provinces was crushed. In 1873 Baker returned to Cairo with a record of successful work behind him, which must have astonished no one more than the Khedive himself.

But once the strong hand was removed, the stone which had been heaved uphill with so much labour rolled swiftly down again. Less than a year elapsed between the departure of Baker and the arrival of Colonel Gordon, on his appointment as Governor-General of Equatoria. Even in that short time the Egyptian occupation had become merely nominal. Two posts only were held, Gondokoro and Fatiko. Three large slave-trading stations were in full swing on the Bahr el Zeraf alone, whilst on the Bahr el Ghazal the notorious Zubehr had established himself as a practically independent potentate, and was even preparing on his own account an invasion of Darfur. The situation called forth Gordon's fullest energies.

Never did he perform better work than during his three years in Equatoria. As far as it could be done under Egyptian supremacy, he checked the slave-trade and laid the foundations of good government. The country was organized and divided into districts with proper garrisons both along the Sobat and the White Nile. The tribes were peaceful and contented. Communication was established with the great lakes; Lake Albert Nyanza was for the first time circumnavigated. A treaty was made with M'tesa, King of Uganda, recognising his independence, and Emin Bey was sent to represent Egypt at his Court. In 1876 Gordon returned to England.

In Egypt, meantime, the Khedive's reckless extravagance was fast hurrying him to disaster. But the more involved he became, the more he extended his ambitions, like a ruined spendthrift who must keep up his credit at any cost. Extension of his Empire became a mania. After Equatoria came the turn of Darfur in 1874. Darfur had maintained its independence for over 400 years under an unbroken line of Sultans. One of them, Abd-el-Rahman the Just, had entered into correspondence with Napoleon during his occupation of Egypt, and congratulated him upon his defeat of the Mamelukes. Napoleon replied in a remarkable letter:

'*To the Sultan of Darfur*, 12 *Messidor, Year VII. In the Name of God, compassionate and merciful; there is no other God but God! To the Sultan of Darfur, Abd-el-Rahman.*

'I have received your letter, and understand its purport. When your caravan arrived I was absent in Syria punishing and destroying my enemies. I pray you send me by the first caravan 2,000 black slaves, over sixteen years of age, healthy and strong. I will buy them from you. Order your caravan to come immediately, without delay. I have given orders for its protection all along the route.

'BONAPARTE,
'*Commander-in-Chief.*'

Many motives combined to make Ismail desire the annexation of the country. There were long-standing frontier grievances. Its commerce and slave-trade were still considerable, and Ismail hoped to profit by the one as well as to suppress the other. The copper-mines of Hofrat-en-Nahas in Southern Darfur were also a powerful attraction in view of his failing treasury. They were reported to be extraordinarily rich, with veins standing 2 feet out of the ground. And, since Zubehr could not be prevented from his proposed expedition, the only course seemed to be to join him in the conquest. Accordingly, an expedition was despatched from the north, while Zubehr co-operated from the south. The Sultan and two of his sons were killed in battle, and another troublesome province was added to the Khedive's dominions. Zubehr was made a Pasha, but was refused the governorship, which he claimed as his right. For a moment he seemed inclined to assert his independence, but in the end he rashly determined to press his case in person at Cairo, leaving his son Suleiman to fill his place in his absence.

The same year saw the annexation of Harrar at the request of the Mohammedans in that country. Raouf Pasha, who had been left in charge at Gondokoro after Baker's departure, was sent for that purpose. He showed that he had learnt but little wisdom or mercy from his association with Baker and Gordon, for he commenced his administration by strangling the late Sultan, a wholly unnecessary act. Egyptian territory had been extending down the Red Sea coast for some years previously, at the expense of Abyssinia. Little was gained, except some friction with the King of Abyssinia, for the occupation was generally quite nominal, and the tributes imposed by Egypt were seldom or never paid. Still, Egypt now possessed the coast-line as far as the Straits of Bab-el-Mandeb, and the purchase of the port of Zeila from Turkey in 1875 extended her possessions to Cape Guardafui and beyond, and consolidated her position, such as it was, in Harrar, and what is now Somaliland. But even this did not satisfy the Khedive.

Gordon had clearly perceived that it was impossible to properly control the Equatorial provinces without an outlet to the Indian Ocean, to replace the long and precarious line of communication by the Nile. He therefore advised the Khedive to occupy Mombasa from the sea, while he himself co-operated from inland by way of the Victoria Nyanza. Although the difficult nature of the country and the inferior quality of his troops soon convinced Gordon that it was impossible for him to carry out his part in the project, a naval expedition, known as the Juba River Expedition, set out in 1875 under the command of McKillop Pasha. Missing the mouth of the Juba River, which had been selected by the Khedive as their objective, without any great knowledge of the geography of these regions, they ran further south, but encroached on the territory of the Sultan of Zanzibar. At the instance of Great Britain the expedition was given up, but no objection was raised to the recognition of the Khedive's authority as far south as the tenth degree of latitude.

It was not long before the folly of all this extravagant expansion began to appear. Abyssinia had had many causes of quarrel with Egypt, and, in fact, the relations between the two countries had consisted of a state of intermittent war or at best armed neutrality since the time of Mehemet Ali. But the purchase of Zeila thoroughly alarmed King John, and he took steps to protect his rights. War followed, disastrous to Egypt. A first army under Arendrup, a Dane, was, after a few preliminary successes, totally annihilated. A second was hastily fitted out and despatched. It was a model of all an army should not be, and had not King John been diverted by the fear of his rebellious vassal, Menelik, it would no doubt have suffered the fate of the first. As it was, it managed to return to Massowah after two or three months' tedious and inglorious campaigning and one severe defeat.

The falling house had been extended in every direction, but it was falling still. The Abyssinian trouble was not the only sign of weakness. The internal condition of the provinces was as bad as ever; the whole country was restless and disturbed; the feebleness of the Government was increased by the huge distances over which its garrisons were scattered; the slave-trade, openly favoured by the officials whose duty it was to suppress it, was recovering from the blows which had been dealt it. One more heroic effort was made to set things right. Gordon was recalled to Egypt and made Governor-General of the whole Soudan, including Equatoria and the Red Sea provinces. He was specially charged with the duty of settling matters with Abyssinia, of suppressing the slave-trade, and of improving communications.

It was with the utmost reluctance that Gordon returned to the Soudan. Under existing conditions his task was a hopeless one, and he knew it. His letters show how clearly he foresaw the coming catastrophe. He was aware that, while the Egyptian Government was concluding an anti-slavery convention with England, it was at the same time intriguing almost openly with Zubehr, the prince of slave-dealers. Less than half-hearted support was to be expected from Cairo. At the very moment that he needed them most, great part of his troops were withdrawn to serve in the Russo-Turkish War. Immediately on assuming the reins of government he found himself confronted by a serious rebellion in Darfur; at the same time the slave-dealers in the south were gathering in force under the leadership of Suleiman, Zubehr's son. But Gordon faced the peril undismayed. His activity was almost superhuman. Hastily patching up a temporary arrangement on the Abyssinian frontier, he travelled to Khartoum by way of Kassala, Gedaref, and Sennar. Here he spent some time in carrying out a number of reforms, and in May set out to deal with the rebellion in Darfur. Darfur had had its first experience of Egyptian rule, and Harûn, a member of the deposed royal house, had taken advantage of the general discontent. There flocked also to his standard the nomad Arab tribes, who had stood aloof when Darfur was first conquered, but now saw their hereditary trade threatened. But Gordon in a few months had drawn together his scattered garrisons and scattered the rebels by a display of force. By the almost unsupported exercise of his personal influence he next dispersed the combination of slave-dealers under Suleiman, and by the end of the year he was back again at Khartoum, after another visit to the Abyssinian frontier and a round by Suakin and Berber.

Early in the next year he made the tour of all the Red Sea provinces, including Harrar. But in spite of all this ceaseless activity, the close of the year 1878 found things worse than ever. Trouble was once more stirring on the Abyssinian frontier. The northern provinces were quiet, but their quiet was the torpor of exhaustion. In Darfur Harûn had once more reappeared, and in Kordofan, too, a rising had taken place. In the Bahr el Ghazal Suleiman had this time revolted in good earnest. Only Equatoria was comparatively happy under the rule of Emin. Unyoro had been given up, and the Somerset Nile was now the southern boundary. Gessi had been despatched by Gordon to the Bahr el Ghazal, and in March he himself undertook an expedition to Kordofan and Darfur, with the object of pacifying the country and preventing help being sent to Suleiman. At the same time he made great efforts to put down the slave-trade, and captured many caravans. In June he was able to return to Khartoum, and left for Cairo to confer with the new Khedive Tewfik. The short remainder of his time as Governor-General was occupied with his mission to Abyssinia, and in December he left for England.

Meantime Gessi had been performing marvels in the Bahr el Ghazal. Gessi is one of the heroic figures of Soudan history. An Italian by birth, he served in the Crimea as interpreter to the British troops. In 1874 he joined Gordon in Central Africa. Thanks to his energy, a dangerous rising of the Shilluks at Fashoda was put down, and it was he who circumnavigated Lake Albert. The rising he had now to face was most serious. Zubehr, before he left for Cairo after the conquest of Darfur, had assembled his officers and made them swear to revolt when he should send them word that the time was ripe. That time was now come. Their organization was complete; their following was numerous and well-equipped; they had already proved the incapacity of the Government troops. It seemed that there was nothing to stop them from dividing the Soudan provinces among themselves, and finally throwing off the yoke of the hated and despised Egyptian. They even announced that they intended to seize Egypt itself. But they had reckoned without Gessi. With a handful of troops, inferior in quality as well as in number, short of supplies and ammunition, impeded by floods and generally cut off from his base by the sudd, in two notable campaigns he utterly broke their power, seized and shot Suleiman and his principal confederates, and completely liberated the Bahr el Ghazal.

His end was characteristic of the fate of those who, in those dark days, dared to serve the best interests of Egypt. In spite of his successful administration of his province, he found his position made impossible for him after Gordon's departure. In September, 1880, he resigned. After nearly perishing in the sudd, which blocked his steamers, he reached Khartoum. But there, under the new regime, he was no welcome guest. He struggled on to Suez, where he soon after died, unrecognised and unrewarded, utterly worn out by his exertions and privations. When such a man met with such treatment, it was clear that there was no chance of the regeneration of the Soudan under Egyptian rule. The last ray of hope had been extinguished with Gordon's departure.

CHAPTER XV

THE PAST—*Continued*

GORDON'S successor at Khartoum was Raouf Pasha, the very man whom he had dismissed from his post as Governor of Harrar two years before for oppression and other malpractices. The appointment was not of good augury, and the improvements which had been made in the administration fell rapidly to pieces. The slave-traders began to lift their heads; once more the caravans, with their miserable human freight, began to journey towards Egypt or the Red Sea; once more the horde of tax-collectors felt themselves set free to levy their exactions at their will. The critical position of affairs in Egypt itself was reflected in the Soudan. Economy and retrenchment were the order of the day; under such a Government they were not likely to mean anything else than an increase in the burden on the provinces. Colonel Stewart, in his report on the Soudan, published in 1883, graphically describes the character of the irregular Bashi-Bazouks who were employed to collect the taxes.

'Many, if not most of these men,' he says, 'are very indifferent characters. They are mostly swaggering bullies, robbing, plundering, and ill-treating the people with impunity. Probably for every pound that reaches the Treasury they rob an equal amount from the people. They are a constant menace to public tranquillity, and before any amelioration can be expected they must be got rid of. As soldiers they are valueless, having no discipline, nor, except in talk, do they exhibit any extraordinary courage. Compared with that of negroes and Egyptians their pay is high.'

A last paper reorganization took place in 1882. Schools were to be set up, a proper system of justice established, the slave-trade was to be again suppressed. A capable Governor-General, Abd-el-Kader, who might have made a great mark in happier times, was appointed to supersede Raouf. But it was too late. The promised reforms could be no more than acknowledgments of past deficiencies. Egypt had wasted the golden opportunities created for her by Baker and Gordon. The day of retribution was at hand. From every quarter of the horizon the clouds had long been gathering, and the tempest was now to burst. In August, 1881, the Mahdi had proclaimed himself publicly during the Feast of Ramadan as the prophet foretold by Mahomet.

Sheikh Mohammed Ahmed, strangely enough the son of a carpenter, and a little over thirty years of age, was a native of Dongola, but had been living for some time in the neighbourhood of Khartoum, where he had acquired a great reputation for sanctity. The Soudan was then, and still is, a soil peculiarly suitable for the growth of prophets. The creed of Islam, as is well known, has a peculiar fascination for the natives of Africa, owing, perhaps, to the simplicity and directness of its essential formula. But it has generally only overlaid, and not displaced, the old superstitions of pagan days. The people are naturally credulous and excitable, and in the entire absence of any education or learning there has been nothing to restrain them. Almost any holy man can gain some kind of a following—at any rate for a time. But the Mahdi was more than an ordinary prophet, and the times were ripe for rebellion.

The venality and corruption of the Egyptian officials had produced widespread discontent among the agricultural and trading classes. The oppressive exactions of the tax-collectors had reduced many to ruin and beggary. It seemed that any change of rulers must be a relief. The spasmodic efforts to suppress the slave-trade had seriously alarmed the nomad tribes, whose chief source of wealth it was. The one thing wanted was some strong influence to bind together the discordant elements of revolt. It was found in the idea preached by the Mahdi of the regeneration of Islam by force of arms. After the long years of wrong and oppression the promise of universal equality, with one law and one religion for all, was sweet to hear. To the people that sat in darkness it seemed that a great light had indeed arisen. The four great sects of the Mohammedan Soudan sank their differences in favour of the Prophet of the new dispensation. The stern character of his teaching, his earnest denunciation of earthly vanities and pleasures, his prohibition of the use of intoxicants and smoking, did much at the outset to rouse the fanatical fervour of his adherents. It was a social as well as a religious revolution. The simple jibbeh, or cotton shirt, became the badge of his followers and the emblem of his doctrine.

The course adopted by the Egyptian authorities was well calculated to increase his prestige. First they ignored, then they underrated, his power. Feeble attempts were made to seize him, which failed utterly. At first Sennar was the centre of disturbance; a little later the Mahdi appeared in Southern Kordofan. A small force of regulars was sent against him in December, 1881; they fell into an ambush and were utterly destroyed. In June, 1882, Yusuf Pasha, Governor of Fashoda, with several thousand men, set out to crush him finally. Yusuf had distinguished himself in Gessi's famous campaign. That an experienced leader of disciplined troops could have anything to fear from a crowd of half-naked Arabs seemed to him utterly impossible; so he neglected every precaution, with the inevitable result of annihilation. Throughout the Soudan it was regarded as little less than a miracle. The warlike tribes of the west began to flock to the Mahdi's standard. His fame spread to the farthest extremities of Darfur. By the end of 1882 he had three armies in the field. Everywhere the Egyptian garrisons were beleaguered. The hated tax-collectors and outlying bodies of troops were the first to feel the vengeance of the rebels. Early in 1883 came the fall of El Obeid, after a most gallant resistance under Mohamed Pasha Said. The Egyptian Government made a desperate attempt to regain their position. Hicks Pasha arrived at Khartoum with 10,000 men, mostly raw recruits, and advanced into Kordofan. Betrayed by their guides, suffering terribly from heat and thirst, deficient alike in courage and discipline, the expedition was beaten before it set out. On November 5 the massacre, for it was nothing else, took place, and only 300 survivors were left. The effect was tremendous. On the one side it was decided to abandon the Soudan and withdraw all the garrisons, and Gordon was sent to carry out this task; on the other, a gigantic impulse was given to the cause of Mahdism.

Great as Gordon's influence had once been, it was impossible for it to outweigh the religious fervour of the Mahdists. His position was immensely weakened when it became known that his final object was the evacuation of the Soudan. Many waverers felt that they must in the end be left at the mercy of the Mahdi, and wavered no more. Gordon's view was that the only way to evacuate the Soudan with honour and safety was to hold on to Khartoum, crush the enemy if possible, and then to retire, leaving behind some form of government capable of maintaining order. The only man, in his opinion, capable of such a position was Zubehr. But his views were not shared, perhaps hardly understood, by the authorities at home. The British Government, committed sorely against their will to a military occupation of Egypt, and earnestly desiring to make it as short as possible, were most unwilling to undertake further responsibilities in the Soudan. The conflict of opinion produced paralysis. Neither policy was completely adopted. British troops moved up to Assouan. In the Eastern

Soudan the disaster of El Teb, where once more the Egyptian troops under Valentine Baker allowed themselves to be slaughtered like sheep, and the fall of Sinkat, forced the Government to take military action. General Graham, starting from Suakin, won brilliant victories at El Teb and Tamai, and relieved the garrison at Tokar. But the proposal to make a dash on Berber, now in the hands of the Mahdists, was vetoed, and Graham's troops retired. Meanwhile in the Nile Valley the golden hours were flying fast. Gordon stood firmly by his policy of holding Khartoum, partly in the hope, growing daily fainter, of being able to turn the scale by his own personal exertions and influence, partly with the intention of forcing the British Government to change what seemed to him their disastrous and dishonourable policy of abandoning the Soudan to the rebels. On their side, the Government drifted, hesitating and temporizing, reluctant to reverse their settled policy of peace. And so the tragic game was played. Month by month the investment of Khartoum grew closer. Slatin Bey in Darfur, after fighting twenty-seven battles in the course of 1883, was deserted by his troops and forced to surrender. El Fasher was reduced in the following January. In the Bahr el Ghazal, Lupton Bey, the successor of Gessi, after holding out for eighteen months, shared the same fate, and was brought in a prisoner to Omdurman. In the east the garrison of Gedaref made terms with the enemy. The Mahdi was able to concentrate his forces at the important point.

Only in August, 1884, a relief expedition was

decided on, and at the end of December two columns started from Korti, one along the Nile towards Abu Hamed, the other across the desert by Gakdul Wells to Metemmeh and Khartoum. But the effort that might have succeeded six months earlier was too late. Two days before the relieving steamers arrived the weak defences of Khartoum had been stormed, and Gordon had fallen.

For the moment it seemed that Gordon's death would do what he himself could not do when alive. Great preparations were made for a new campaign in the autumn. The columns, which had drawn back to Merowe after much severe fighting, were quartered for the summer along the river. General Graham was again despatched to Suakin with orders to crush Osman Digna, and a railway from that place to Berber was begun. For two months there was hard fighting at Hashin, at McNeill's Zeriba, and at Tamai. But the hot fit passed. The railway, which had been begun without any survey, quite in the familiar Egyptian fashion, was given up. The whole Nile Valley was left unoccupied as far as Kosheh. The dervishes pushed on, and occupied Dongola, and, though decisively beaten at Ginnis at the end of the year, they continued to maintain a harassing border warfare.

The Mahdi died in June, 1885, carried off by a malignant fever, or, as some say, poisoned by a woman, whilst in the middle of his preparations for the invasion of Egypt. His successor, the Khalifa Abdullahi, a Baggara of the Taaisha tribe from Darfur, took up his plans. To the children of the desert Egypt might well seem an easy and attractive field for plunder. They had had a rough experience of the quality of British troops in action, but these new antagonists had always followed up their victories by retreat and evacuation. They had no reason to doubt the continuation of the same nerveless policy. The British Government talked loudly of abandoning Egypt itself. Any resistance by Egyptian troops unsupported seemed an absurdity. But unexpected obstacles arose. A revolt in Darfur, war with Abyssinia, and the opposition of the great Kabbabish tribe, combined to delay the Khalifa's advance. Not until 1889 did he send his best general, Wad el Nejumi, forward. Nejumi pressed on without misgivings. Only Egyptian troops were guarding the frontier. But the Egyptian battalions, trained and led by British officers, were no longer mere droves of frightened sheep. Under Wodehouse at Argin, and under Grenfell at Toski, they first checked, and then annihilated, Nejumi's army. Thenceforward the dervish power steadily declined. The recapture in 1891 of Tokar, Osman Digna's chief base of supplies, put an end to his influence in the Suakin district. Raiding and frontier warfare continued till 1896, but there was no serious fighting, until the Dongola Expedition in 1896 marked the opening of the well-prepared campaign which ended in the complete overthrow of the Khalifa's power.

Once more the day of punishment for the guilty had dawned. In the disasters and defeats inflicted by the Mahdi, in the horrors of the sack of

Khartoum and many another town, Egypt had paid the penalty for her sins against the Soudan. In blood and shame she had reaped a full harvest. But the unhappy Soudanese had only thrown off the yoke of one master to find themselves under that of one much more terrible. Even before the Mahdi's death plunder and success had corrupted the new faith, and dimmed the first glow of religious fervour. The Khalifa, under the guise of religion, had developed into a cruel and bloodthirsty tyrant. He and his principal supporters, Arabs from Western Darfur, regarded themselves as conquerors in a foreign land. Especially since he had brought his own tribe, the Taaisha, to settle in Omdurman, he had made it his policy to depress the power of the Nile Valley tribes like the Jaalin and the Danagla. On them fell the brunt of his military expeditions. Executions, massacre, and confiscations, were the order of the day. The inhabitants of whole districts, especially in the once-populous country along the Blue Nile, were forced to come and live at Omdurman, which grew into a city of 400,000 inhabitants. This concentration, by at once creating a great demand and withdrawing the agricultural population from their proper occupation, produced the most terrible famines. Everything was sacrificed to the supremacy of the western Arabs. In ten years more damage was inflicted upon the country by this pack of human wolves than in all the sixty years of Egyptian dominion, bad as it was. Slatin Pasha estimates that 75 per cent. of the population perished. Great as the abominations of the slave-trade had been before, they were never greater than under the Khalifa's rule. The export of slaves was forbidden, but from all parts of the country they were brought in droves to the market at Omdurman.

The slave-dealing tribes had been the principal instrument of vengeance upon Egypt. It was now their turn for chastisement. The Egyptian Sirdar, Sir H. Kitchener, who fought with the railway as well as with troops, advanced to Dongola in September, 1896, and the banks of the river were again occupied up to Merowe. Next year Abu Hamed and Berber were taken, and the new railway laid from Wadi Halfa across the Korosko desert. Early in 1898 the battle of the Atbara was won by the Anglo-Egyptian troops. In September the great battle of the campaign was fought at Kerreri, Omdurman was taken, and the Egyptian flag was once more flying at Khartoum, this time with the British flag beside it. By the end of the year the authority of the new rulers was established, after some difficult and troublesome campaigning, throughout the Eastern Soudan as far south as Fazokhl and Famaka; gunboats, though blocked in the Bahr el Gebel, had proceeded up the Bahr el Ghazal and hoisted the flags near Meshra el Rek. Before the end of 1899 the Khalifa and the remnant of his supporters still in the field had fought their last fight. In the hour of their downfall the dervish chieftains displayed a splendid courage worthy of their race. Proudly disdaining to be fugitives in the lands they had ruled, they perished to a man with their faces to the foe.

But of far greater importance than the final destruction of the Khalifa were the events that followed on the fall of Omdurman, which are inseparably connected with the name of Fashoda. Since the time when the Khedive Ismail was pursuing his wild career of annexation without attracting any particular attention except from the natives of the countries annexed, the whole position of affairs had entirely altered. In every direction European Powers had appeared upon the scene. In the east Massowah had been handed over to the Italians in 1885, and they had inflicted several severe defeats upon the dervishes. In 1893 a British Protectorate had been declared over Uganda and Unyoro, and in 1895 the British flag was hoisted at Duffile. The Belgians from the Congo Free State had pressed on to the Nile, and in 1897 an expedition under Chaltin was successful in taking Rejaf, which had long been the southernmost dervish post on the river.

Most serious of all was the French movement from the Upper Ubanghi district of French Congoland into the Bahr el Ghazal province. The Belgians had already made considerable expeditions in this quarter, and had penetrated as far north as Hofrat-en-Nahas in Southern Darfur. In 1894 they made over their claims to the French. The French Governor of the Upper Ubanghi, M. Liotard, immediately began to lay his plans with great energy and forethought for the solid establishment of French power in the Valley of the Nile. At the end of 1895 he crossed the Congo-Nile watershed, and seized Tembura on the river Sueh, an affluent of the Bahr el Ghazal. He also occupied Dem Zubehr, on the Bahr el Homr. Captain Marchand arrived with reinforcements from France in 1897, and after spending some time in consolidating his position by occupying posts throughout the country, he set out on the final stage of his journey. After great difficulties and some fighting with the dervishes, he reached Fashoda just eight weeks before the fall of Omdurman.

The French preparations had not escaped the attention of those responsible for Egypt. As far as the Khalifa was concerned, the advance of the Anglo-British troops might have been delayed for some time; but in view of the French advance it was absolutely necessary for Egypt to reassert her rights in the Soudan emphatically and at once. Immediately after the occupation of Omdurman Lord Kitchener hastened on to Fashoda, and found Marchand already there. The situation was grave. For France it was galling in the extreme to be foiled just in the moment of success, but for Egypt the question was vital. The stake at issue was not the possession of a few acres of swamp, but the control of the summer water-supply. Marchand's mission was by no means the mere freak of an adventurous traveller, anxious to hoist his country's flag. It was undertaken as part of a policy skilfully planned and deliberately pursued.

It was no mere coincidence that in the previous year the Bonchamps Expedition had set out from Abyssinia, and endeavoured, though vainly, to join hands with Marchand from the east. Firmly based in the Bahr el Ghazal, masters of the Upper Nile Valley, and joining hands with Abyssinia, the French would have been in the end complete masters of the fate of Egypt. If the French had insisted, there must have been war. Happily, they gave way, and by the agreement of 1899 withdrew all their posts in the Bahr el Ghazal. The boundary of the Soudan was fixed along the Nile-Congo watershed.

Thus the Soudan emerged at last, and finally, it may be hoped, from her ordeal of blood and fire. Her history will be no longer a record of tyranny, rebellion, war, and famine, but of steady progress under a just and civilized government. The work of Baker and Gordon is bearing fruit at last. Great as were the joy and relief with which the downfall of the dervish tyranny was hailed throughout the Soudan, it was largely due to faith in the word and just dealing of Englishmen—a faith that was first established by them and those who worked with them—that the whole country settled down so quickly.

Among all the pioneers of good government in the Soudan one name stands out conspicuously. Most of the ease and success with which the present government works is due to the realization of the reforms recommended in Colonel D. H. Stewart's Report on the Soudan. The rulers of the Soudan are the first to acknowledge their obligation to that masterly document, an epitome of keen observation and practical wisdom. Fortunate is the country that is served by such men as he. A few months after he had completed his report he returned with Gordon to Khartoum. He, at least, had no delusions as to their prospects of success. And yet he went as cheerfully and lightly as though

> "Tendens Venafranos in agros
> Aut Lacedæmonium Tarentum."

To die treacherously murdered on the Nile bank was a sorry fate for such a man. The road to the Soudan is strewn with the bones of many victims. But of them all it may be said, and especially of him: Their work lives after them, and their memory is nobly avenged.

CHAPTER XVI

THE NEW KHARTOUM

No one whose lot it has been to travel through the night on the plains of a tropical country can forget the amazing effect produced when the sun, with one great leap, as it seems, springs clear above the horizon. All in a moment the world's face is altered. Its features are the same, yet utterly changed. The traveller, however hardened, can scarcely fail to wonder at the transformation. A journey to the Soudan to-day produces a very similar impression on one whose mind is full of the memories of the dark past. The din of battle has hardly ceased to echo, but the transformation is complete.

Even after Egypt, with all its fascinations, rich with the remains of ages of civilization, full to the brim of questions and problems deeply interesting to the student of history, archæology, politics, or economics, the Soudan, with its triple capital, Khartoum, Omdurman, Halfaya, comes upon you with a freshness and charm that are indescribable. Travelling by the ordinary methods, you may go from Alexandria to Khartoum in about six days. It is well worth while, even for anyone who has been up and down the whole length of Egypt, to take the whole journey in one piece. There is all the excitement of starting for a new country, and at the same time an opportunity to gather into a focus all the old impressions. Easily and smoothly you swing through the fertile cotton-fields of the Delta, and its populous cities and villages, prosperous but dirty, and at Cairo you settle down into a most comfortable sleeping-car for the night journey to Luxor.

Early next morning you are in the cane-fields of Upper Egypt, with the river close on one side and the desert on the other. At Luxor you must change on to the narrow gauge for Assouan, and there is time to refresh yourself with bath and breakfast, and to look across at the Plain of Thebes and the Valley of the Tombs of the Kings, or to ride a donkey out to Karnak. From Luxor to Assouan it is hot and dusty enough, and you are glad to rest there for the night. Next day you embark at Shellal, above the Dam, for Wadi Halfa, a voyage of some 200 miles. Coming down, the steamers do it in about twenty-four hours, but up-stream it is a leisurely voyage of three days. There is plenty of time to see the interesting antiquities of Nubia, and, above all, the famous rock-hewn temple of Abu Simbel, the colossal statues of which are, perhaps, the most impressive of all the monuments of Egypt. It is, besides, a most beautiful reach of the river; the hills come down to the water in bold and rugged outlines, showing to perfection in the pure, dry, desert air.

The effect of the Dam is clearly seen as far as Korosko. First of all, at Shellal the boat is moored amid a grove of palm-trees, the temples of Philæ are knee-deep in water, and the Nubian villages look quaint enough as they stand on the edge of the desert, forlornly mourning their strip of cultivated land, most of which the greedy Reservoir has swallowed up. Probably a great part of these people will now migrate to Dongola, but the loss of the land—for which, indeed, compensation has been paid—is really a small matter to them. Hardly anywhere is an able-bodied male to be seen; all are away making their living as sailors or servants elsewhere, leaving the women and old men to keep their homes. These Nubian boatmen are a most happy and thrifty people, ready to work all day and dance all night, always to the accompaniment of a song.

The boundary between Egypt and the Soudan, settled by the Convention of 1899, runs along the twenty-second parallel; not far beyond this is the frontier town of Halfa. There is no mistaking the signs of British rule. The whole place is rigidly clean, an extraordinary contrast to the filth of the Egyptian villages. The streets are well laid out and scrupulously swept, and shady avenues of trees are springing up. But at present Halfa is not particularly interesting, except as the railway terminus of the Soudan. It is twenty-eight hours to Khartoum. Nothing can be more comfortable than the well-appointed sleeping-car train, which runs twice a week. Starting at eight in the evening, you strike right across the Nubian Desert, most desolate and forlorn of countries. The very stations have no names, but are known merely by their numbers.

In the morning you come to Abu Hamed, back

to the Nile once more. Abu Hamed is just at the elbow of the river, where it turns to the west for its great circuit by Merowe and Dongola. Here was the scene of one of the stiffest fights in the Soudan Campaign, when General Hunter made his dash from Korti, in 1897, further down the Nile, to seize the point for which the new railway was making from Wadi Halfa, and here are the graves of Fitzclarence and Sidney, officers of the 10th Soudanese, who fell in the battle. Around this spot a ghostly legend hangs. It happened that the other white officers of the battalion were wounded on the same day, and the black troops marched back to their bivouac without any of their white leaders. A black regiment is always accompanied by its women on the march, and these have high notions of military honour. They would have nothing to do with men who dared to return alive from the field on which their officers had fallen. The warriors quailed before their wives, and serious trouble was brewing, till a black sergeant, who lay dying of his wounds, solved the difficulty. 'Tell the women,' he said, 'that enough of us are dead to guard the spirits of the white men in the other world. I myself will mount the guard.' There are innumerable witnesses to testify that he has kept his word. The lesson was not lost. It was the same battalion which later, at the Atbara, raced the Camerons for the enemy's zareba, and, catching their Colonel as he ran in front of them, bore him heels foremost right through the camp, securely hedged by a living wall of bodies, because a second loss like that of Abu Hamed was not to be thought of.

From here onward the journey is full of interest. Berber is springing up again from its ruins; it even boasts two stations, but it has not an attractive look as a place to live in; there is as yet nothing more than the mud huts of the country, and it is the hottest place in the world. Next comes the Atbara River, though not the scene of the battle, for that was thirty miles upstream; then Shendi, of fiery memory, but now the Crewe of the Soudan, and finally, late at night, you step out of the train at Halfaya, the railway terminus. One glance at the sky will show you that you are really in the tropics. Canopus is shining fiercely in the east. Right overhead the giant Orion strides across the vault. Northwards the Great Bear stands like a huge note of interrogation in the sky, and just over the opposite horizon the Southern Cross is looming up.

Halfaya stands on the northern bank of the Blue Nile, near its junction with the White. It is destined to be the workshop and commercial quarter of the capital. At present the Government steamboat factories and workshops are still at Omdurman, a legacy from dervish days. But already the necessary buildings are springing up, and as trade increases the railway terminus will attract more and more to its immediate vicinity. The Blue Nile, just before its junction, divides into two channels, which embrace the fertile island of Tuti. Opposite Halfaya, and along the southern bank of the island, the river runs in a glorious sweep due east and west for two or three miles. Here, facing northwards, stands Khartoum, with as imposing a situation as any capital could wish for.

A well-made road runs all along the river-front, which is being gradually embanked and walled. Right in the centre rises the white Palace, the official residence of the Governor-General, a handsome building, on the site of Gordon's old palace, set in a lovely garden. On either side of it stretch a succession of Government offices and the neat residences of Government officials, to the new, spacious, and comfortable hotel on the one side and the Gordon College and British barracks on the other, pleasantly variegated with gardens and groves of palm-trees, acacias, limes, and bananas. Behind this Government belt, the town is carefully laid out into wide streets and squares in two other belts. The second of these contains, or will contain, houses and shops built by private persons, but of a good class and on approved plans; the third is open for the erection of any buildings that the owners choose to construct. Finally, close to Gordon's rampart are the Soudanese barracks, and, right outside, the native villages, laid out in squares allotted to different tribes, where you may see huts of every shape, characteristic of many different parts of Africa.

Considering that three years ago Khartoum was nothing more than a dirty dust-heap, the work that has been accomplished by the Royal Engineers is truly wonderful. Of course, the city is still in the hands of the builders; everywhere are gangs of workmen levelling roads, preparing foundations, making bricks of Nile mud, carrying, hammering, digging, building. Women, too, are employed. Their principal duty is bearing water for the lines of young trees that will one day make each street a shady avenue. Already the town is lighted with lamps far better than many an Egyptian city, and it is hoped that in a short time a tramway will be in working order.

From November to March the climate is very delightful: it is hot, of course, at times, but the north wind blows steadily and coolly practically every day, and sometimes the nights are even cold. Even if it is hot, there is always the Blue Nile to refresh the eyes. It is a comfort to find that the Blue Nile is really blue—as blue as any Italian lake. One is so often told that it is called blue because of the mud it brings down during the flood—the mud which causes the 'red' water so dear to the Egyptian cultivator. But anyone who looks at it cannot fail to realize that its name is derived from its clear and limpid waters, and not from its muddy flood-time. One of the most charming scenes in Khartoum is the view of these blue waters seen from the windows of the Soudan Club through a green maze of palms and lime-trees.

Khartoum, with its bungalows, offices, shops, and banks, is a civilized town, summoned up out of nothing, as it were, by an enchanter's wand. Far different is Omdurman. Here, too, the engineers have been at work, clearing, demolishing, and cleaning. Only those who marched in after the battle four years ago and saw those foul labyrinths of streets can realize how much has been accomplished. But though purified and greatly shrunk, of course, within its limits in the Khalifa's time, Omdurman remains a real Central African city, with nothing European about it. It was originally intended to move all the inhabitants over to

Khartoum; but the natural convenience of its position on the left bank of the river, just at the junction of the two Niles, makes it impossible to carry out this intention. Its population is actually increasing at present, and it seems far better to let natural forces work their way, and retain it as the native quarter of the capital, distinct from the seat of government. It possesses, too, a wide sloping foreshore, or beach, exactly suited to the feluccas which ply upon the river.

This beach is one of the great sights of Omdurman, and a fascinating spectacle it is. On a busy day it is absolutely crowded with traders and porters from all parts of the Soudan. It is the great market-place for gum from Kordofan, feathers from Darfur, ivory from the Bahr el Ghazal, dhurra from the Blue Nile country, and everything else that comes in by boat or camel. All day long the porters go to and fro, carrying their loads and chanting their monotonous songs, chiefly tall, broad-chested, but spindle-shanked negroes from the Nile Valley, Dinkas, Shilluks, Bongos, or Bari, with here and there a short, thick-set, sturdy hillman from Southern Kordofan or Dar Nuba. Women equally diverse in type sit sorting the different qualities of gum. Naked children, brown and black, tumble and chatter in every direction. It is difficult to drag one's self away, so strange and novel and varied are the sights.

But Omdurman has much more to show. First and foremost the Khalifa's house, the only two-storied building in the town, and built of brick. It is occupied now by the British inspectors.

Hard by are the ruins of the Mahdi's tomb, too solidly built to be entirely destroyed, but even its partial demolition has been sufficient to put a complete stop to pilgrimages. In front is the great square in which the Khalifa used to preach to his assembled dervishes. It is surrounded by a high wall excellently built, said to be the work of the German Neufeld. It is witness now of scenes very different from those of the old fanatical days, so far removed in everything but time—perhaps the parade of a Soudanese battalion, a football match, or the arrival of a string of camels laden with gum from El Obeid.

In the Beit el Amana are some interesting relics of the past. Here are the brass cannon taken originally by the Mahdi from the Egyptians, and piles of captured dervish muskets of every shape and form, swords and caps; and here, too, is the Khalifa's carriage, a ramshackle, broken-down old four-wheeled chariot. It is a puzzle how it ever got to Omdurman. It looks as if it were built for some quiet old maiden lady in a French provincial town, but it fell on a strange master in its old age; and Slatin Pasha, now Inspector-General of the Soudan, had to run as a footman before it. Close by is the great broad street leading northwards, by which the British troops marched in on the the battle. Every step is reminiscent of the last evening of days of the dervish tyranny.

But the most fascinating sight of all is the sook, or market. The mixture of races is amazing. It would take a trained scientist to catalogue them all. From a camel to a silver bracelet, there is nothing dear to a native that cannot be bought.

Here are made and sold the angarebs, or native bedsteads, woven with string across a low four-legged wooden framework. As yet these people have not much mechanical ingenuity, and their appliances for working in wood and iron or other metals are rude in the extreme. The past in the Soudan has not been conducive to the development of the arts, but there is promise for the future.

One feature of the place is the curiosity shops, full of many strange spears and shields, knives and sabres, and occasionally an iron gauntlet or coat of mail such as the dervishes wore; but these are rare now. These Arab shopmen know how to drive a hard bargain, and love nothing better than chaffering over a price. All curiosities are called 'antiquos.' There is an annual flower-show in Khartoum, at which a prize is given for the best 'antiquos.' A Soudanese turned up who was very anxious to compete for this prize with a live porcupine, which he insisted was an 'antiquo.' He was allowed to exhibit at last, but not for competition, on condition that he himself led it about by a string, as the authorities could not undertake its charge. With more justice a live tortoise is also commonly offered for sale as a very choice 'antiquo.' In another quarter are the rope-makers, twisting their long coils of fibre in exactly the same way as they have been twisted for centuries; further on, the basket and net makers are weaving dried reeds in curious patterns. Yet another part is entirely devoted to the needs of women. Here are combs and strings of all kinds of beads, unguents, and the strange preparations they use for fixing the small plaits in which they tie their hair; cinnamon bark and other scents, without which no wedding is complete; cooking-pots and other articles of their scanty domestic furniture; even dolls of most weird and fantastic shapes and material, and many other quaint trifles. The whole place is crowded with women, black and brown, many of them tall in stature, easy and graceful in their movements, and, however ugly they may be in other respects, nearly all with the most beautifully shaped arms.

It is a strange contrast to go a little further on through the western outskirts of the town, and to find a game of polo proceeding on the hard desert sand. And beyond, again, in the middle of these great spaces, far from the rush and turmoil of the town, are the quiet graves of two or three of those who have laid down their lives in this far country, and shall see their homes no more.

It would take a long time to get tired of merely riding about in Omdurman and watching the thousand and one sights of such a place, and reflecting on the stranger scenes that must so frequently have been enacted there less than five short years ago. But if you want a gallop in the desert, nothing is more delightful than to pass along the broad street leading northwards through the deserted quarters of the once huge city, and to ride, with the keen wind blowing freshly in your face, to Gebel Surgham, the hill which overlooks the field of Kerreri, that historic field of bloodshed. The merest novice can easily follow every phase of the battle, and see where wave after wave of the dervish hosts rushed madly but heroically to their doom. Some of their skulls still lie bleaching in the sun. There, too, is the khor where the Lancers made their famous charge. Or you may take boat, and sail past the mud forts that saluted the steamers which came just too late for the watcher anxiously straining his eyes to see them from the palace in Khartoum. Almost every spot has its own historic interest. Perhaps as you skim along before the wind a battered old paddle-steamer labours creaking past, towing a string of barges. It is one of Gordon's gunboats, once more patched up and restored to duty.

But of all the sights and interests of this fascinating place, by far the most impressive, as in Cairo itself, is the ancient and mighty river. Khartoum and Omdurman are what they are because here the two great tributaries join their forces and set out across the waterless desert on their great mission to Egypt. The spot where the Blue and White rivers meet, and for some distance flow side by side unmingled, would be still in many ways the most notable in all Northern Africa, even if Khartoum and all its eventful history were blotted out.

CHAPTER XVII

THE NEW SOUDAN

KHARTOUM would be a sad place to visit if it were nothing more than a city of memories. Happily, it is no longer so. It has indeed a history behind it, full of lessons and warnings which cannot be ignored. But the Khartoum of to-day is looking forwards, and not backwards; it is the young capital of a young country. It is barely four years since the final defeat and death of the Khalifa, and it is only from that date that the establishment of settled government can be reckoned.

The Sirdar, Sir Reginald Wingate (the post of Governor-General of the Soudan is still combined with the command of the Egyptian Army) might well have despaired of the task which confronted him of administering those vast regions, 1,000,000 square miles in extent, when he succeeded to his post. The difficulties to be faced were enormous. It seemed as if every germ of civilization had been extinguished by the dervish rule. Everywhere villages were deserted; lands, once fertile, had gone out of cultivation; actual famine was seriously threatened. The decrease of population was extraordinary. Whole tribes enumerated in Colonel

Stewart's list had totally disappeared. When the British troops reached Metemmeh in the final march to Khartoum, they found the place a shambles. The once powerful tribe of the Jaalin had been massacred almost to a man by the Baggara dervishes. That was but one instance out of many of the Khalifa's methods of government.

From a country so wasted and desolate it seemed hopeless to expect a revenue for many years. The fabric of administration had to be built up again from the very bottom, and there was no money to do it with, far less to provide for that capital expenditure always necessary in a new country. Only Suakin, Halfa, and Dongola, from their longer military occupation, possessed even the rudiments of the machinery of government. No money was to be expected from England. More than that, every British officer in the Soudan was turning longing eyes to the war in South Africa. It was difficult to get new men with such an attraction elsewhere. Yet in spite of this it may be fairly said that the Soudan government has come safely through the troublous period of infancy. Sheer hard work on wise and statesmanlike lines has had its due effect, and the British officials, most of them soldiers whose abilities would have brought them to distinction elsewhere could they have been spared, can so far congratulate themselves on the result of their patient labours.

Though greatly shrunk within its former limits, the Anglo-Egyptian Soudan is still of very great extent. On the north the boundary with Egypt is now the twenty-second parallel, following this line, which passes just north of Wadi Halfa, right across the desert to the Tripoli border on the west, and to the Red Sea on the east. It then follows the coast past Suakin as far as a point about seventy miles south of Tokar, where it meets the Italian colony of Eritrea. Here, turning inland, it runs south-west, crossing the river Gash just above Kassala, down to a point on the Setit River, another tributary of the Atbara. Here Abyssinian territory begins, and the boundary trends more to the south than before, going by Gallabat and crossing the upper waters of the Rahad and the Dinder, tributaries of the Blue Nile, till it reaches the Blue Nile itself just above Famaka and Fazokhl. Hence it runs right south to the west of the Beni Shangul hills, across the Baro, Pibor, and Akobo rivers, all in the Upper Sobat district, till it reaches a point where the sixth degree of north latitude cuts the thirty-fifth degree of longitude east of Greenwich. The sixth parallel is roughly the line of division between the Soudan and the Uganda Protectorate as far as the Nile, the southernmost Soudanese post on the Nile being at Mongalla, and the northernmost British at Gondokoro. On the other side of the Nile the boundary, starting from Lake Albert, runs north-west for a very long distance along the Nile-Congo watershed, first along the Congo Free State, then along French Congoland, till it reaches the borders of Darfur. The corner, however, between the Nile and this line, as far west as the thirtieth degree of longitude east, and as far north as a point nearly opposite Mongalla, is known as the Lado Enclave, and is still held by the Belgians, being leased by them during the life of King Leopold. On reaching Darfur the boundary runs northwards along the western edge of that country to about the thirteenth parallel, and thence northwest across the desert to the borders of Tripoli.

In this vast area, some 1,200 miles in length and 1,000 in breadth at the broadest part, there is naturally an immense variety of country. The Northern Soudan is, with the exception of a strip of land along the Nile, almost a desert; it is, in fact, a lesser Egypt. Up to Shendi it is a rainless country. From Shendi to Khartoum, and some way south, there is a regular but not extraordinary rainfall during the months of July, August, and September. Further south, about parallel 13°, tropical rains begin, becoming heavier and longer towards the Abyssinian hills on the one side and the Upper White Nile and the Bahr el Ghazal on the other. With the rainfall the character of the country changes. To the west of the White Nile, between parallels 15° and 11°—that is, nearly as far south as Fashoda—the country of Kordofan and Darfur stretches in vast plains or steppes, covered with low thorny trees, mimosa and gum trees, and prickly-grass. Water is scarce, and stored in wells and the trunks of baobab-trees. There are occasional hills, which become greater and more numerous towards the west and the south. In Southern Kordofan the plains and valleys between the hills are rich in vegetation and huge trees. They are impassable during the rains, and also after them, until the long grass has been cleared by burning.

The Bahr el Ghazal is a real land of waters,

the Punjab of the Soudan, for it has five principal rivers, all affluents of the Bahr el Ghazal. Part of the country consists of little but the most dismal swamps; but nearer to the watershed there is much rolling country between the rivers. The soil is fertile, and there is an abundance of forests and parklands. To the east, along the course of the Atbara, and between the Blue and White Niles, there are vast level plains of alluvial soil, very fertile when flooded or after the rains. The country gradually rises towards the Abyssinian frontier. At wide intervals great masses of rock protrude, some consisting of excellent granite, better than that in the Assouan quarries. As the country rises it becomes more and more wooded, until there is dense jungle, especially near the Blue Nile. The scenery all along the Abyssinian frontier is described as very beautiful and mountainous; but most of it is very unhealthy, especially during and after the rains, when the climate is fearfully hot and oppressive, and malarial fevers of several kinds are common even among the natives themselves. The country south of the Sobat near the Nile belongs to the swamp districts. Further inland is the least explored and most uninhabited part of the Soudan; it is not likely to attract much attention at present.

The natives of the Soudan fall roughly into two divisions—Arab and negro. Generally speaking, the Arabs are all north of a line drawn east and west some distance below Fashoda, the negroes south of it. But any precise classification of races is impossible. The Arabs, as a conquering race, imposed their religion and their language upon the original inhabitants. Many so-called Arab tribes are really almost negro. The confusion of blood is very great. Slavery has introduced a very strong negro element into many tribes whose Arabic descent is unquestioned. Sometimes, when the original race was not negro, they have preserved their own language. Thus, the people of Dongola speak a tongue of their own besides Arabic, and several of the nomad tribes in the Halfa-Suakin district are apparently not Arabic in descent.

The inhabitants of Kordofan afford an example of the mixture of blood. The sedentary village population consists of the aboriginal inhabitants, with an admixture from Darfur. Under the Egyptian conquest a good deal of Turkish and Levantine blood was introduced, as well as negro blood from the South by means of slavery. The nomad tribes are a much superior race mentally and physically, and talk a much purer Arabic. Their great occupations are cattle-breeding, carried on by the Baggara tribes, and camel-breeding, carried on by the Kabbabish and the Beni Gerrar. Besides these there are also the hill tribes, negroes, very black in colour and small in stature, peaceful by nature, but warlike when roused by the attacks of the nomad Arabs. Of all the negro races in the Soudan, these are the most intelligent and hard-working.

The whole of the Soudan is now directly governed, with the exception of Darfur, by the Governor-General and his subordinates. Darfur, though a tributary State, once more enjoys internal independence under its own Sultan. After the fall of Omdurman Ali Dinar, a member of the former ruling house, then resident in Egypt, was sent back to Darfur. The people had been for some time uneasy under the Khalifa's hand, and he found little difficulty in establishing his position. His prestige and power were considerably increased in 1902 by the surrender to him of the last remnant of the dervishes, who had fled westwards from Bor, on the Nile, their last station. The remainder of the country is divided into provinces, or mudiriehs, for the purposes of administration. The capital, Khartoum, with its sister towns and a district within a radius of ten miles, forms one province by itself. Immediately south of it is the Ghezireh (the Island), comprising all the rich and fertile district between the two Niles, formerly known as the island of Sennar. It is the kernel of the Soudan. South and south-east of the Ghezireh, stretching along the Abyssinian frontier, and extending from the White Nile on one side across the Blue Nile to Gallabat and Gedaref, is the province of Sennar. North of Sennar, along the upper waters of the Atbara and the boundary of Eritrea is Kassala. Suakin is the maritime province. Halfa is the northern frontier district. Dongola and Berber extend along the Nile. Kordofan, largest of all in extent, occupies all the country west of the White Nile as far as Darfur. All these make up the Arab and Mohammedan Soudan. The pagan negro portion is divided into Fashoda Province, including all the Sobat country, and the Bahr el Ghazal Province, to the east of the Nile and south of Kordofan.

Each province is ruled by a Governor, or Mudir, a British officer of the Egyptian Army ; under him are one or two inspectors, also British officers, and one or two subinspectors, who are English civilians, members of the newly-formed Soudan Civil Service. The mudiriehs are subdivided into mamuriehs, or police districts, presided over by a mamur, or inferior magistrate; all the mamurs are Egyptian officers. Each province is, in many respects, a little empire by itself. Where the distances are so great, there must necessarily be a great deal of decentralization, and every Governor has a great deal of independence. But he is, of course, subject to the authority of the Governor-General, and his expenditure is carefully controlled by the central Finance Department.

The first duty of the provincial Governors is to maintain peace within their borders, and to make the inhabitants acknowledge the authority of the Government. In the elder provinces, Halfa, Suakin, Dongola, and Berber, there was no difficulty. The nomad tribes, the only possible source of trouble, are regularly paying the tribute, which has been imposed on them more as a badge of authority than for revenue purposes. As long as this is paid and they behave well, they are left to the rule of their own sheikhs, according to their custom. At Khartoum the presence of a British battalion insures order among a people who have not forgotten the sound of bullets. The stories of mutiny among the black troops were greatly exaggerated. It lasted only for a day, and was nothing more than a skirmish between two battalions which had a difference over some women. On the borders of Darfur there was a slight disturbance, but Ali Dinar is a strong and capable ruler, who understands very well that it is his interest to be on good terms with the Government, and the matter has been satisfactorily settled. In South-Western Kordofan the Nubas in the hills have been accustomed for centuries to raid each other, and be raided in their turn by the Arabs of the plains. In such remote regions it is difficult to summarily suppress customs of this sort, but it is hoped that permanent peace will soon be established by means of camel-corps patrols. The same applies to the frontier raids on the Abyssinian side. There was some reluctance at first to deal drastically with these raiders, for fear of trouble with Menelik, but now that the boundary has been definitely settled by the recent agreement, the raiders can be punished without any question of violating the frontier. In Fashoda Province the tribes, even the wild and shy Nuers, are settling down quietly, and though they hate the 'Turk,' they are glad to welcome the British.

The Mek or King of the Shilluks, who live on the banks of the Nile from Lake No to a point above Fashoda, gave some little trouble at one time. He was very much struck with the institution of tax-collectors, which he determined to imitate. Accordingly, he put some of his people into a kind of uniform, and sent them round the villages to collect women and corn, nominally for the authorities of Khartoum. But he was informed that, while the Government admired his zeal for civilization, they felt, at the same time, that his knowledge of it was so elementary that it was necessary for him to devote some time to further study before he put his theory into practice. He is now undergoing a course of tuition at Wadi Halfa.

The Bahr el Ghazal is the most recently occupied province. Posts are now occupied at Wau, Rumbek, Dem Zubehr, Shambe, Chak Chak, Tonj, Meshra-el-Rek, and Channamin on the Jur River. The country was occupied without difficulty. Here the two principal tribes were the Dinkas in the north and east and the Niam-Niams or Azande in the south. All the smaller tribes, Jur, Bongo, Golo, etc., had been broken up by repeated raids, and had fled to one or other of these powerful neighbours for protection, the protection practically taking the form of slavery. A tract of 150 miles square had thus relapsed into absolute wilderness, though once thickly peopled. Now the tribes from the Dinka side are returning and rebuilding their villages under British protection, an immense advantage to the country, for they are hard-working and industrious, while the Dinkas are lazy and troublesome in civil life, although good soldiers. Early in 1902 one of the Dinka tribes fell upon a convoy and murdered a British officer near Rumbek by treachery. But the murder was speedily avenged, and this was the only instance of active hostility. The Niam-Niams, although cannibals, are much more intelligent and progressive. They are very well disposed and anxious to trade. The worst enemy in the Bahr el Ghazal is not any human foe, but disease, especially the blackwater fever, which has been of a severe type, and more than one valuable life has been sacrificed to it. Fortunately, as experience of the country grows, the danger diminishes, and it is hoped that as the country is developed, with better housing and communication, and less moving about in the rains, still greater improvement may be looked for.

On the whole, the Soudan is far more peaceful internally than it has ever been before. Nothing has had a more pacifying effect than the spectacle constantly witnessed of British officials roaming the country unarmed and unescorted. This visible confidence has done much more than years of campaigning could ever do. There have been, it is true, one or two occasions when troops have had to be employed. The Soudan is still full of combustible materials. Fanaticism and ignorant superstition combined are dangerous elements. But the Government has taken to heart the lesson taught by the rise of the Mahdi. When troops are sent, care is taken that they shall strike swiftly and effectively. A new Mahdi appears in the Soudan once or twice a year. Once a woman proclaimed herself, and gained a considerable following. Sometimes they die away of themselves, but all require to be carefully watched. Such an one a few months ago had to be put down by an expedition in Kordofan. His capture and defeat at once put an end to his divine pretensions. These petty outbursts are legacies from the unhappy past; under good government they will become less frequent and less dangerous. Every day the Pax Britannica grows stronger and more deeply rooted in every direction. Peace once established, the British officer finds his real work only begun. The restoration of order was comparatively simple, the kind of business for which he had been trained. His regimental duties may have given him some slight acquaintance with accounts and book-keeping, or the elements of military law and procedure. Every scrap of such knowledge must be beaten out and made to reach as far as possible, and so guide him in his financial and judicial duties. But where will he have learnt to make bricks, to measure land, to make roads, to pronounce on methods of gathering rubber or gum, or the diseases of cattle or dhurra? Yet all these things he will have to do, and many others in the course of his busy days. One of the great attractions of the Soudan is the splendid keenness of all these men, soldiers and civilians alike. From Wadi Halfa onwards the same spirit is observable in everyone you meet. Grumbling you will hear, but it is the grumbling of keen men. The Governor of one province is convinced that too much is being spent on the capital. If only some of that money were given to him, he would soon show startling developments. Another thinks his trade suffers because the railways or the steamers are not so good as they ought to be. Another is dissatisfied because he cannot get a schoolmaster sent him, and so on. All are not, of course, equally competent (that could not be), but there are no idlers. All are working heartily for the good of the Soudan and its people, and the honour of their own country. Many of the Governors and Inspectors of Provinces have fought in command of the black battalions against the Arabs. They learned to like the one and to respect the other. The liking and the respect are repaid tenfold. There could be no better foundation for sound government, and from Suakin to El Obeid, and from Halfa to Gondokoro the Englishman is obeyed and trusted as no ruler of the Soudan has ever been before. It is, perhaps, hardly necessary to add that all these men are young. The Sirdar himself, old in experience, is young in years, and from him downwards, whether on the civil or the military side, there is hardly a man over forty. The administration of these vast regions in a tropical climate is no sinecure. A young country needs youth and activity in its rulers.

It must not be forgotten that we have been immensely assisted in our task by the Egyptians. All the minor administrative posts are held by them. Most of the clerks in the various Government offices are Syrians or Copts from Egypt. Without this small fry of officialdom, it would have been difficult, if not impossible, to carry on at all. Many of these Egyptians have done and are doing their work faithfully and well, but it has been a hard struggle to fight against the old Soudan traditions. In former days, when a man went from Egypt to serve in the Soudan, he expected either to die or to get rich in a very short time. Death was always before his eyes, so he made haste to be rich while he could. To the fellah the Soudan meant nothing less than death; his family mourned him as lost, and lost he generally was; but the official, if he came back at all, was set up for life. Of course, any case of taking a bribe or extortion is severely punished whenever detected; but, in spite of this, corruption undoubtedly does go on to a certain extent, so much is it in harmony with the traditions of the official and the expectations of his victim. Moreover, the Egyptian is not very capable of enduring a bad climate. He seems to give up heart immediately. A class of natives educated enough to take his place in the Government service is badly required, but does not at present exist.

A great deal is often talked of the evils of military government, and, of course, there comes a time when military methods are out of place. But I doubt very much if any set of civilians could have done so well with a country like the Soudan as the soldiers have. Whatever may be the soldier's fault, what he is told to do, that he will set to work to do, and very likely he will carry it through successfully, however little training he may have had, where a cleverer man might be dismayed by difficulties. In fact, the government is military because it is mainly carried on by soldiers (there was no one else to do it), but it is run on civilian lines. Spend a morning in the office of a Mudir or Governor of a province, and you will go away with an enhanced opinion of the powers of a British officer.

First of all, you will find him talking fluent, if not always perfect, Arabic, that most difficult of tongues. An interpreter is present in case of an emergency, but he is seldom appealed to. In turn, a bewildering variety of subjects is dealt with. A criminal trial has to be steered through one of its stages. Next, two sets of villagers come in with a petition as to a dispute of land-ownership. In a land which has seen such vicissitudes as the Soudan, questions of title are numerous and difficult in the extreme; within the last twenty years the same plot of land may have had several sets of *bonâ fide* occupiers, and finally none. The two parties come in scowling fiercely at each other (they have probably had a fight already on the land in question), and squat down on the floor, while their spokesmen step forward to tell their tale. The Mudir is addressed as the Protector of the Poor, the Shield from Oppression, and a number of other names. When the matter is disposed of, after a great deal of hard swearing on both sides, everybody who has anything to do with it attempts to give his version simultaneously, and finally the whole crowd has to be thrust out separately by the police, still talking as they go.

Then comes a succession of applicants with other grievances and petitions—Greeks alleging some breach of contract, Arab land-owners demanding remission of taxation or loans for a sakieh, or for sowing seeds. Next the plan of a house has to be examined, the quality of certain bricks to be looked into, an auction of river-fisheries to be held, or a contract about a Government ferry adjusted. Then there are questions concerning drains or wells, pumps or police, licenses for boats or liquor, and a thousand and one other questions of administration. Finally, there is correspondence with headquarters, to adjust some financial, legal, or educational matter. It is no use being a Mudir, or, indeed, any other official, under the Soudan Government, unless you are prepared to work like a horse, and deal fearlessly with nearly every department of human life.

CHAPTER XVIII

JUSTICE AND SLAVERY

THERE is no Magna Charta in the Soudan. It is expressly laid down in the Anglo-Egyptian Agreement of 1899 that the country is to be governed under Martial Law, and all legislative power resides in the Governor-General, subject to the approval of H.M.'s representative in Egypt. But Martial Law is a vague and unsatisfactory term. During the recent operations in the Transvaal a flying column was sent to occupy a railway terminus and to secure the rolling-stock therein. The seizure was successfully accomplished, but when the engines were examined they were all found to be defective in certain essential parts of the mechanism. The commander of the column sent for the railway officials and ordered them to restore the missing pieces. In vain they protested that they knew nothing of them. The soldier, an Australian, was a man of few words. 'I shall shoot you in the morning,' said he, 'unless the pieces are found.' They objected that such a course was utterly illegal, even under Martial Law. 'My view of Martial Law,' he replied, 'is that you must do what I tell you or take the consequences. In any case, discussion is useless. The shooting-party is ordered at eight.' The engines were complete in every particular long before the fatal hour; but such methods, though sometimes necessary in war, are not suitable to peaceful administration. Martial Law remains a power in the background, but the Government lost no time in setting up a system of Civil Justice, and one of its earliest acts was to appoint a properly qualified Legal Adviser.

It is fortunate that English barristers are accustomed to give advice on a great variety of matters. The Legal Adviser, especially at first, when he acted temporarily as Director of Education, had very multifarious duties. He was at that time the only judge in the country, and, besides strictly legal matters, had to assist in drafting regulations relating to customs and commerce, game, travellers, and many other subjects. But his chief occupation was the preparation of penal and civil codes. With one exception, which will presently be noticed, there was no existing foundation to build upon. Under the old Egyptian rule the criminal law supposed to be administered was the Khanoon Humayoun, a Turkish code dating from 1837, but this was perfectly unsuitable for use by a civilized Power. It was wisely decided to make use of Indian and Colonial experience. A penal code and code of criminal procedure were promulgated as soon as possible, founded on the Indian codes, modified to suit the simpler requirements of the Soudan, and made to resemble courts-martial in their procedure to facilitate their working by soldiers. This was followed by a civil ordinance, likewise adapted from the Indian Civil Procedure Code, as used in Burma and British Bechuanaland. By the end of 1901 a system of civil justice was at work in all the provinces except Kordofan, Fashoda, and the Bahr el Ghazal. It has since been extended to Kordofan, and even in the Bahr el Ghazal and Fashoda a beginning has been made.

In each province the inferior police-court work is done by the Egyptian mamurs, who exercise summary jurisdiction on a small scale. All the more important cases are heard by the Mudir, inspectors and sub-inspectors, who form various courts of differing strength according to the nature of the trial. Besides the Legal Adviser there have also been appointed three English judges, all trained English barristers. Their headquarters are at Khartoum, where they form the superior court and court of appeal, and they also travel on circuit in the provinces, either to relieve a particular pressure of work or to avoid the trouble and expense of bringing up cases to Khartoum. At present, of course, any very definite or precise division of duties among the different members of the judiciary is impossible. The ordinary judicial business has been immensely added to by the work of Land Commissions, and every available person has had to be pressed into the service. Naturally, after twenty years of anarchy, of emigrations and immigrations, and of general unsettlement, the ownership of the land is a matter of the greatest possible doubt and confusion, and endless disputes arise. In the early part of 1903 one of the English judges was sitting in Dongola Province on a Land Commission. His preliminary list alone contained 400 cases; and Dongola is by no means the most contentious province. But this work will now soon be completed, and a more complete organization of the courts will be possible.

Meanwhile the criminal law has been very well administered, and, if the complete inexperience in such matters of the British officers who have had to act as judges in the provinces is taken into account, the administration of the civil law has also been most creditable. Mistakes are made, of course, but substantial justice has been done. Confidence in the tribunals, as well as the need for them, is growing rapidly. The fact that the amount of work done in the civil courts in 1902 was exactly double what it was in 1901 is striking testimony to the growing belief in their impartiality.

The only department in which the whole system had not to be built up again from the bottom was the administration of the Mekhemeh Sharia, the courts dealing with the sacred Mohammedan law. Just as in our own country the ecclesiastical courts formerly dealt with all questions relating to wills or marriages according to the canon law, so these courts administer the sacred law in all matters relating to succession, marriage, personal status, and charitable endowments. Once they covered the whole field, but as the sacred law was unchanging, or nearly so, other codes had to be introduced to suit changing conditions, and their action was gradually restricted to the subjects mentioned. In the old Soudan nearly every place of any pretensions had one of these courts presided over by a Cadi, and they doubtless dealt with many matters outside the limit, for civil cases could be tried by the Cadi, if both parties were agreed. The Cadi was frequently paid a very small salary, or none at all, and was expected to make his living out of court-fees and the presents he received, a system by no means conducive to strict justice. Now these courts are much fewer in number than formerly, but each Cadi receives a regular salary, and all the fees go to the Government. All the Cadis are under the direct supervision of the Grand Cadi or Cadi of Cadis, assisted by a Mohammedan inspector. It was very difficult at first to find properly qualified occupants for these posts. Among native-born Soudanese there were practically none, and the supply had to come from Egypt. If they were honest, it was at any rate something to be thankful for. But in days to come, especially when the Training College at Khartoum has been more developed, it may be hoped that they will be not only honest, but also capable, and that they will also possess a sound knowledge of the law which they have to administer. It is worthy of note that a Soudanese Cadi has been appointed at El Obeid.

The sacred Mohammedan law is a very delicate matter for Europeans to deal with. It abounds in defects, but suggestions for reform coming from a Christian would not be well received. There are also opportunities for conflict with the ordinary civil courts. For example, in a question of inheritance a dispute as to whether the deceased person was a Mohammedan or not could only be properly settled by the civil court. But all difficulties have been avoided up to now by tact shown on both sides. The Grand Cadi is a man of singular enlightenment and ability, and is working hard to improve the administration of the Mohammedan courts. His remarks on the general situation, as quoted in Lord Cromer's Report for 1902, are worth repeating:

'What consoles me with regard to the stagnant state of the Mohammedan law in this country is the fact—deplorable though that fact be—that this state of stagnation is general throughout the Mohammedan world. And, although, for many centuries, it has been the only law applied to the people, time has, on account of the failure of those in charge to administer it properly, necessitated the introduction of other codes.

'The whole responsibility for this decline in the Mohammedan law must be attributed to the authorities of former times.

'What renders this condition of things the more regrettable is the fact that some of the causes which have for long been undermining the stability of this law have become a part of its traditions, any digression from which would be considered as a deviation from the Sharia law. Hence, it is impossible for me to exaggerate the difficulties which obstruct the way of the reformer. Yet I hope that we may have a good opportunity in this country to improve this state of things, and to bring about an unprecedented epoch of advance in the history of reform.'

Apart from the defects of the religious law, there are many obstacles to a proper administration of justice in the Soudan. Justice is a new idea. Colloquial Soudan-Arabic has no very extended vocabulary. A British officer learning Arabic at Khartoum had impressed upon his teacher that he only wished to master the ordinary language of the country. One day a new word turned up. 'Is that a good word?' he asked his teacher. 'Good?' said the learned man. 'I should think so, indeed. Why, if you use that word, you and I will be the only people in the Soudan who understand what you mean.' The word might well have been the ordinary Arabic term for justice. There has seldom been any use for it until now in common parlance.

To the mind of the Arab the notion of an impartial tribunal giving final decisions is an absolute novelty. His natural view is that the judge decides either according to his own caprice, or according to the greater bribe, or to please some great man. The image of blind Justice holding equal scales is very puzzling to him. A judge, therefore, who, when he has decided a case, thinks he has heard the last of it, is liable to rude disappointment. The unsuccessful suitor is very likely to reappear a month or two later and ask for at least a modification of the judgment. He knows now that a bribe is worse than useless; so he comes with a terrible tale of ruin and despair to move the compassion of the Father of the Oppressed and the Protector of the Poor; or—for he is full of resources—he alleges that the judgment has not been properly carried out; or that at the trial an important witness on his side was absent; or that some third party, who had an interest in the case, never heard anything about the trial at all. It may be all pure invention, and very often is, for the Arab can be a fertile liar—that difficulty is not peculiar to the Soudan—but part of it may be true, and has to be sifted. It is very probable that some interested party will not know the date of a trial, when distances are so great and publication impossible other than by word of mouth, in view of the universal illiteracy. So the decided case has to be gone into once more. In the present stage of civilization it is often better to use the wisdom of Solomon than the wisdom of the code, and justice must necessarily be more or less patriarchal, for it is of far more importance to gain the confidence of the people than to abide by the strict laws of procedure.

So, too, in administering the criminal law the magistrate finds himself sometimes confronted by difficulties which could never be contemplated by any legal code. Alike in the Mohammedan and the Pagan Soudan ancient beliefs and superstitions live on, fostered to an incredible extent by the backward state of the people and the prevailing ignorance. In the face of the habits of mind produced by these conditions the principles of jurisprudence lose their significance. The ordinary rules cannot be justly applied to such cases. Not long ago a native of Southern Sennar was brought to trial on a charge of murder. There was no dispute as to the facts. The victim had been savagely done to death in cold blood. The accused admitted the deed, but pleaded, in all good faith, a defence which seemed perfectly natural and satisfactory to himself and every other native. His brother had recently died, and he had ascertained that his death was due to the evil eye. In accordance with the moral code of the district, it became his duty towards his brother to exact vengeance, and he therefore killed the man whose evil eye had caused the mischief. In his view and that of all his neighbours there was no more criminal intent in the act than if he had killed a poisonous snake.

Another most interesting case is related by the Legal Adviser as having recently occurred in Dongola Province. Taha Ali and Ahmed Hamad carried on business in partnership as butchers, and Taha kept the purse. One day Taha told his partner that ten and a half dollars belonging to the partnership had been stolen. But Ahmed did not believe him, and roundly accused him of stealing the money himself. A violent dispute arose, but at last they agreed to refer the matter to a holy man then residing in the neighbourhood. This holy man was a fakir belonging to Timbuctoo, who had made the pilgrimage to Mecca, and was now returning. At first he was very reluctant to interfere; it was no business of his, he said; they should go before the mamur. But the more he refused the more eagerly they insisted. They even said that if he would not, it was because he could not. It was a poor sort of fakir that could not find out a little thing like that, and they would lose no time in proclaiming the fact. At last he consented to act. First he copied out some passages from some religious books which he had with him on a native writing-board with European copying-ink. Then he washed off the writing into a bowl with bread and water, and divided the bread and water between the two, telling them that the one in the wrong would shortly become very ill. Each man consumed his portion of the mess and went away. An hour or two later Taha Ali was seized with violent pains in the stomach, and, returning to the fakir, confessed that it was he who had stolen the money. But in spite of his confession the pains grew worse, and he died the same day. The authorities stepped in, and the fakir was put on his trial for poisoning. The facts of the case were undisputed, and the fakir himself gave evidence on his own behalf. He pointed out that he had only undertaken to investigate the affair very reluctantly, as he was well aware that it was not his business, but he could not damage his reputation as a holy man. He had adopted the best method of discovering the truth. The man's death was the act of God, not his. As for the suggestion that the copying-ink was poisonous, he was willing on the spot to drink up the remainder of it, the bottle being still about half full. The medical examination revealed no sign of poisoning, and the case was referred to Khartoum.

With the view of discovering the drift of the best native opinion on these subjects, the Legal Adviser told the story to two natives, one a religious sheikh of very high position, universally respected, and the other a servant who had been for many years in the employment of English masters. The sheikh, whilst not doubting that such an ordeal, if employed by a man of holy life, was a reasonable method of detecting crime, was inclined to think that this particular fakir was an impostor. At the same time, he did not consider that he should be punished, as the death might be due to some cause hitherto undiscovered. To illustrate his point he repeated a well-known story of a man who died at his friend's house immediately after eating some honey. Great suspicion fell upon the friend, who only escaped punishment by the discovery of a dead serpent coiled up at the bottom of the pot. In this case, too, he suggested, a snake might have spat into the inkpot. The servant went further. He, too, was of opinion that trial by ordeal was a reasonable method of detecting crime, and more than that, it was really the only satisfactory and effective way, far better than any investigation by the best and wisest of mamurs. The only thing that surprised him in the story was that the guilty man should have died after confessing his crime; for this was contrary to precedent. He could only conclude that the man was exceptionally wicked, and that God had taken this opportunity to punish him for other crimes.

In such an atmosphere it is no wonder that miracles abound and holy men thrive. It is exceedingly difficult to know how to deal with them. Like the magicians of Ancient Egypt, whose descendants they are, they are sometimes open to the suspicion of establishing their miraculous reputation by natural but very undesirable methods. At Berber the Mudir was anxious to embellish the place with avenues of trees. So he imported some libbek acacias from Egypt, and to insure their being watered announced that rewards would be given to anyone in front of whose house a libbek was planted as soon as it attained a certain size. For a time all went well, and the trees grew and flourished. But then a local fakir saw his chance. He proclaimed that watering trees was contrary to the will of God, and threatened the most terrible penalties on anyone who dared to disregard his orders. Only one man, a sergeant in a black regiment, was bold enough to flout the fakir, and to continue watering his tree. Within a short time the man himself, his wife, and his servant, were all dead. It may have been a coincidence; it may have been the effect of imagination acting on uncultured minds; but far more probably, though it could not be proved, it was due to some wicked contrivance of the holy man. In the case of the Dongola fakir, a very practical solution of the difficulty was adopted. He was not punished, but facilities were given for his return to Timbuctoo. Guilty or not, a man whose reputation requires such bolstering up is a very undesirable resident.

That the various judges and magistrates administer the law intelligently and with discretion, and that the people themselves are more and more contented with the law, and accept it even when it comes in conflict with old-established ideas, is shown by the decrease in the number of petitions to the Governor-General. In the Soudan, as in other Oriental countries, anybody who has a grievance is allowed to appeal directly to the highest authority, and as the Governor-General does a great deal of travelling every year, there is every facility for presenting them. At first they numbered several thousands every year, but in 1902 they sank to the comparatively small total of 600. But whilst every endeavour is made to govern the country on lines acceptable to and understood by the people, there are, of course, some points on which the policy of a civilized government is necessarily in opposition to very deep-rooted customs and habits to such a degree as to completely upset the old basis of social life. In a minor degree this is true of the partial application of the game laws to natives, but of far greater importance is our attitude towards the institution of slavery and the slave-trade.

The occupation of the Soudan has been a tremendous blow to slavery; one of the principal recruiting-grounds for slaves has practically been closed. A certain amount of slave-raiding goes on along the Abyssinian frontier. Descents are periodically made by parties one to two hundred strong, well armed, from the south-western districts of Abyssinia. They raid the Barun negroes, and carry off the women and children. The same kind of thing is apt to happen on the Darfur frontier, and some of the remote tribes in the same quarter sometimes raid each other with the object of getting slaves. Some of these, but not many, find their way to Dongola, or the Ghezireh; others are taken to Tripoli. Special steps have now been taken by the Anti-Slavery Department of the Egyptian Government, which now has its headquarters at Khartoum, to put down this traffic. Two extra English inspectors have been posted, one at Rosaires, on the Blue Nile, the other at El Obeid. They are to form small mounted corps of the best Arabs and patrol the disturbed districts.

Apart from the actual work done, it will be a great thing to enlist the best of the Arab tribesmen in the Government service, and it is hoped that they will in time form the nucleus of an effective native police. These men have been in the past some of the principal exponents of slave-catching themselves; they ought to be very adept at their new business. With these exceptions the slave-trade within the borders of the Soudan has practically disappeared. During the first year or two captures were occasionally made of small caravans, but very heavy penalties were imposed. There is still a constant demand for slaves in Arabia, and once a slave is shipped over the Red Sea a good profit is assured. But it is too dangerous now for anyone to try to make a regular livelihood by it. There are still, however, about twenty cases a year of trials for offences against the slavery laws, mostly isolated cases of kidnapping a woman or a child, and probably there is besides a fair proportion of undetected cases. But the regular trade is pretty well stamped out.

The benefits of the abolition of slave-raiding and kidnapping are immediate and obvious to everyone. Even the Arab can understand them, but he finds it very difficult to appreciate our attitude—which, needless to say, is uncompromising enough—towards slavery as a domestic institution. It is the one serious complaint which he has against the new government. His domestic habits and customs have been completely based on slavery for centuries. Slavery is permitted and recognised by the Koran. In most cases the slaves themselves have been treated more like members of the family than as slaves, and no doubt many of them have had far happier lives than they would have had in their own villages. Nor is it difficult to point to evils which have arisen from the emancipation of the slaves. It is a melancholy fact that many of the towns in the Soudan are crowded with freed slaves, too lazy to do anything but steal, while the women have recourse to an even less reputable occupation. It is easier to break down the social system of centuries than to build up a sounder fabric in its place. But the thing had absolutely to be done if the Soudan was to have a real regeneration. Even when the slaves have been well treated, the demoralization caused by slavery has been great. The Arabs have all the vices of a slave-owning people. It was a good time to make an absolutely fresh start. All changes of such magnitude are bound to produce dislocations. The evils of the change will die out with the present generation. The good must be waited for patiently, but it is sure to come.

CHAPTER XIX

EDUCATION AND THE GORDON COLLEGE

PROBABLY the last thing that a military government might be expected to take an interest in is education, and yet few educational establishments are so widely famous as the Gordon College at Khartoum. As yet it owes its reputation to the fact that it was founded by one great soldier in memory of another, not to its achievements ; it was born great. But already it has justified its founders ; its mere existence marks an extraordinary contrast between the character of the present régime in the Soudan and that of any preceding.

Whenever the Soudan is mentioned, the first question asked is, 'How is the Gordon College getting on?' and the question cannot be answered in a word. The actual building is indeed for the present complete. It is a handsome structure of native red brick, built in the Moorish style, but retaining the collegiate character. It occupies two sides of a square, the front facing on the river. In the centre is the principal entrance, and over it a tower. If the original design is finally carried out, the whole quadrangle will be completed. Along the inside runs a cool and airy cloister, with winding stairs leading to the upper story; the class-rooms are spaciously designed. Its commanding position at the east end of the town makes it a conspicuous landmark for many miles round. From no point is this so remarkable as from the hill of Surgham, which overlooks the battlefield of Kerreri. Here is summed up much of the past and the future of the Soudan. On the one hand is the scene of the final overthrow of the forces of darkness and ignorance by war; on the other the symbols of that longer contest for the conquest of the Soudan by the peaceful arts of science and learning.

With the eye of faith it is easy to look forward into the future, and to imagine the time, generations hence, when the Gordon College will be a true centre of learning for all these vast territories. Then it will stand, a completed quadrangle, in the middle of large gardens, its own territory, as green and cultivated as they are now arid and dusty. Its halls and class-rooms will be crowded with picked students from all the provincial centres, not vainly pursuing a dry and vain scholasticism, as in the other degenerate Universities of the Mohammedan East, but eagerly following in the paths of living science, and learning by practical teaching in the laboratory and workshop to wrest from Nature her secrets, and to absorb the principles underlying practice in the departments of chemistry and medicine, mechanics, agriculture, and the arts. Perhaps once more, in years to come, the culture and science of the Arabs will be as famous as they were in the great days of Arab dominion.

It is a long way to travel from such stimulating forecasts to the actual state of learning and education in the Soudan to-day. There was not much learning under the Egyptians, but at least a certain amount of theological study went on. Under the Khalifa even that was sedulously discouraged, and the books were ordered to be destroyed because he feared that they might tend to discredit the unorthodox doctrine of Mahdism. Reading and writing were not likely to flourish under a ruler who, possessing neither of these arts himself, and entertaining strong suspicions of those who did, was wont to give drastic expression to his views. As a consequence, there never was a country more absolutely and wholly illiterate. Writing is practically an unknown art, and reading hardly less so. It is perfectly useless to post a Government Proclamation unless a competent person is stationed by it to read it out to any passer-by. At the same time, there flourishes the most exaggerated respect for a written document, which is regarded as a kind of magic book, and cases have been known in which swindlers have extorted large sums of money by going round exhibiting a paper professing to be an order to pay issued by the Government. Obviously, education has had to be on very humble lines at first, and must continue so for some time.

Was, then, the Gordon College a too ambitious attempt to anticipate the future? Is it a mere white elephant, doomed to be a vain monument of an ill-directed wave of enthusiasm? Such a view is far from the truth. It would be strange indeed if a project so dear to the heart of Lord Kitchener was of such a nature. Certainly, it is impossible to start a complete University right away with a building and an endowment of some £4,000 a year. Time is of the essence of the question. It is possible to argue that the money used in the building might have been more advantageously expended in other ways. But, apart from the fact that the subscribers doubtless wished to see some immediate result for their munificence, I am sure that it was the right policy to build at once.

For the Gordon College, though not yet a University, is much more than a college. It is the centre of all the new educational and intellectual influences in the Soudan. Its director is also head of the Education Department; the activities of both are inseparably connected. It acts as an extraordinary stimulus upon the authorities in the direction of education. It would have been so easy and so natural for a Government so hard beset for money to neglect education for other objects, apparently more practical and more immediately pressing. The actual material presence of the college makes it impossible for its claims to be overlooked. Very likely without it there would not have been an Education Department at all. Secondly, but for the existence of the building, the Soudan would certainly never have obtained such valuable gifts as those of Mr. Wellcome's bacteriological laboratory and Sir W. Mather's complete technical workshop apparatus, containing all that is necessary for the establishment and organization of departments for manual training and technical instruction. Thirdly, the building itself has been already, and will be to an increasing extent, of the greatest use; and, moreover, there is still about £100,000 of the original endowment remaining, the income from which is playing a great part, as will be shown, in providing the beginnings of education in the Soudan, and so laying the foundations for the future work of the college itself. Lord Kitchener was wiser than his critics. Among his many claims to fame, none is greater than the clearness with which he saw that a sound educational system is one of the fundamental requirements of the Soudan, as well as a substantial foundation for our rule.

In 1901 the Soudan Government spent £1,421 on education, in addition to the Gordon College endowment; in 1902, £3,577; and in 1903 something over £6,000. With such resources as these, it is obvious that nothing heroic could be attempted. Only the more immediate needs could be attended to. Looking to the necessity for the education of a class of native public servants, it was most important to establish some sound primary schools, in which the boys should be given a fair general education in reading, writing, and arithmetic, besides a certain amount of history and geography and English. Four of these schools are now in existence; two of them, those at Halfa and Suakin, were established some eight or nine years ago by the Egyptian Government, and have only recently been handed over to the Soudan; they are in an efficient condition. The most interesting, considering their recent establishment, are those at Khartoum and Omdurman. That at Omdurman was the first, the direct offspring of the Gordon College. It now numbers over 200 pupils, and is in a most flourishing condition. There are constant applications for admission, partly, no doubt, from the reason that pupils of the school are thought likely to obtain Government employment, but partly also from a real appreciation of the advantages of education.

The school course is divided into four years, and the curriculum is based on the Egyptian one—modified to the extent that no subjects are taught in

English, except English itself. An inspection of the school made it clear to me that, at any rate, in the subject which I could understand—the teaching of English—the methods were thoroughly sound, and the results good. Pupils in their last year are also taught land-measuring. The reason is the great demand which comes from every province for land-measurers—a most important thing in view of the assessments for the land-tax—while no trained men are available. This part of the work is taught both in the class-room and practically in the field. I am certain that in a very short time there will be land-measurers available of very good ability, so excellent was the quality of some of the work done.

Some, of course, of these boys are the sons of Egyptians in the Government service, to whom it is a great blessing to be able to get a good education for their children on the spot. But far the majority of them—at least 90 per cent.—are genuine Soudanese, some of them members of good Arab families, whose fathers were prominent in the service of the Khalifa. Originally it was intended that this school should be transferred to Khartoum, and housed in the college as soon as the general exodus took place. But as Omdurman shows no signs at present of diminishing, and is, indeed, once more increasing, the school has been kept on, and another started on similar lines in Khartoum itself. This school is also flourishing, but it naturally contains a larger proportion of the Egyptian-born pupils; altogether it has about 120 scholars. This school is now housed in the Gordon College itself.

Book-learning is not the only channel of instruction employed. I had the good fortune to be umpire in the first football match between Khartoum and Omdurman schools, in the mosque square at Khartoum. It was a hot afternoon, and I felt as though I should get a sunstroke whilst umpiring; but these boys, all hatless as they were, played with great energy, and appeared to derive nothing but benefit from the heat. They played a good game, and it was pleasant to see that what they lacked in experience they made up in courage and determination. All the players, of whatever shade of black or brown, and the shades were very various, showed a spirit which augurs well for the future. Anybody who can play football with energy in Central Africa must have good stuff in him. The match was drawn.

In another direction also an encouraging start has been made. There is at present in the Soudan no skilled native labour ; blacksmiths, tinsmiths, carpenters, and bricklayers, are all in demand, and have to be imported from outside at great trouble and expense. In the hope of meeting this demand, an industrial school has been established at the Dockyard Works, which are now on the river at Omdurman, but which will eventually be moved to Halfaya. The head of the Steamboat Department agreed to take in sixty boys as apprentices. These are divided into two shifts of thirty each, and they alternately receive a day's schooling and do a day's practical work at their separate trades as carpenters, fitters, or riveters. In school their time is divided between reading, writing, and arithmetic, and also drawing. The plan has been found to answer admirably. Not only does the education they receive improve their intelligence as workers, but some of the boys have shown such proficiency in drawing that they are able to copy engineering and building designs with such accuracy as to be able to relieve the English superintendents of a good deal of work. Many of these boys are sons of men employed in unskilled labour at the works, who take the greatest interest in seeing their sons advance so far beyond themselves. Many applications for admission have to be refused, and there is little doubt that when Sir W. Mather's technical school at the Gordon College is in full swing, it will fill a great need in the requirements of the country.

On the whole, the most difficult task which the Director of Education has to face is that of diffusing the elements of knowledge among the masses of the people. The great distances to be covered alone impose a tremendous obstacle. But it is extremely important that at least a portion of the population should be able to understand the outlines of the machinery of government as laid down in notices and proclamations, so as to be able to protect themselves against the exactions of minor officials and the frauds and deceits practised on them by wandering rogues. It has wisely been determined to proceed along the lines laid down by Mohammedan tradition. The kuttabs, or preliminary schools, are a well-known part of the ordinary religious organization. They are supposed to give instruction in reading and writing and the Koran, and there are many of them scattered over the Soudan, as in other Mohammedan countries. They are, in fact, a sort of private elementary school, something like the old dames' schools once existing in parts of England. Unfortunately, they are almost entirely useless at present. The teachers are incredibly ignorant. What little instruction they give is confined to teaching by rote certain passages from the Koran, the meaning of which is understood by neither pupil nor teacher. The buildings are usually filthy to the last degree. The idea is to establish model kuttabs in different parts of the Soudan, and to make them as efficient as possible, so as to improve the others by their example. To quote the words of the Director :

'The process of formation has been in all cases the same. With the help of the Mudir, a suitable building is put up ; then the least incompetent sheikh that can be procured is installed. After confidence has been established, and the nucleus of a school formed, he is superseded by a trained teacher from Egypt, who, under the local supervision of the Mudir and occasional supervision from my office, begins to reduce chaos to order.'

Progress has not been very rapid. Lack of money and lack of competent schoolmasters sadly hamper all operations. But kuttabs are now established, attached as a sort of junior class to the schools at Khartoum, Omdurman, Halfa, and Suakin. A model kuttab has been established at Berber, which is reported to be doing well, and another is being built at Dongola. The like is also being attempted at Wad Medani, a populous town on the Blue Nile, capital of the province of Sennar, with about 40,000 inhabitants. Reference has been made to the lack of trained schoolmasters. Egypt itself feels this difficulty, and Egypt is at present the only source of supply on which the Soudan can draw. It was to meet this demand—at least, so far as the kuttabs are concerned—that a small training college for native sheikhs was opened in the beginning of 1901 in connection with the school at Omdurman. At first this interesting experiment was not very successful. The students, who all belonged to the best Arab families, were all proud, ignorant, and lazy; and as Arabs they were inclined to despise the Egyptian schoolmasters, whose task it was to teach them. But now there is a great improvement. They have increased in number to about thirty, and only lack of room prevents a further increase. I watched them doing their own lessons, and also receiving practical instruction in teaching by taking a class of the school under the guidance of a master. It was impossible to doubt the value of the experiments. They were nearly all fine-looking, intelligent young men, some of them really handsome, with the keen, clear-cut features that mark the pure-bred Arab. Three of them had come from distant Kassala, where at present there are no means of education whatever. The course lasts three years. At the end of it they are to be examined as to their fitness, and they will then be drafted off either to teach in their kuttabs or else to some posts in connection with the native Courts. Whether as schoolmasters or Cadis, they will be most useful elements in the development of the Soudan.

In the negro portion of the Soudan, inhabited by the pagan tribes, the people are so backward in civilization that the question of education does not at present arise, or, if it does arise, assumes a totally different aspect. Here is the field for the missionary. Two missions are already established—one, the American Medical Mission, on the Sobat; and the other, the Austrian Roman Catholic Mission at Taufikieh, on the White Nile. Both are doing good work, and both are to be encouraged and assisted by the Government. In other parts of the Soudan it must be remembered that we are dealing with a fanatically Mohammedan population, and any suspicion that the Government was trying to proselytize would immediately wreck all schemes of education, and probably be the signal for grave disorders.

It will be a long time before the schools turn out sufficient pupils to fill the Government Civil Service, and there does not seem to be any danger of producing mere 'babus,' hanging about and relying on a certain knowledge of English to procure them a job. The teaching of English is entirely confined to those boys who are going to make use of it in the Government Service or in commercial pursuits, where its knowledge is required. For the ordinary mass of the population nothing is to be gained by an imperfect knowledge of English. The authorities are unquestionably right in discouraging such teaching; the supposed political advantages of it are small, if not entirely imaginary. The Soudan can never be a real white man's country; its rulers must always be speakers of Arabic, and its people will do far better to employ their time in more useful ways than struggling with a foreign language.

The Gordon College is the centre of education, but even now its activities are not confined to mere teaching. It is proposed to form a collection of books dealing with the Soudan, its peoples, its natural history, and its various productions, accompanied by specimens to illustrate them, and some progress has been made. Mr. Wellcome's valuable bacteriological research laboratory is in full working order, and a skilled expert from Scotland has been in charge of it for some months. He is doing a work of great importance, not only to the Soudan, but also to the scientific world in general. The Soudan is a land where strange diseases both of men and animals abound. There is a wide field for research. The scientific and systematic examination of these obscure subjects is already bearing fruit, and cannot fail to ameliorate the conditions of life in these tropical regions for the European as well as for the native.

Lord Cromer has promised that more shall be done for education in the future. There is no need for hurry ; indeed, it is essential that the educational system shall be built up slowly with caution and patience. But in time each province will have its own primary and technical schools, whose pupils will be selected from the elementary kuttabs. Afterwards, when all this has been carefully organized, the provincial schools will in their turn pass on their more promising students to Khartoum to receive the higher education which will then be demanded. Then the Gordon College will at last become a real college. In it the germ of a most hopeful future is contained. If the work proceeds on the same sound lines as hitherto—which there is no reason to doubt—great days are coming. The Arab is capable of a very high degree of civilization, and has a great intelligence, which has as yet had no chance of development. And in those days, I doubt not, those who founded the Gordon College, and first kindled the fire of learning, will be praised by its students as sincerely and as deservedly as the benefactors of any of our own Western foundations.

CHAPTER XX

TRADE AND COMMERCE

Sir Rudolf von Slatin, Inspector-General of the Soudan, who possesses an unrivalled experience of the country, reports that 'the whole situation in the country is very satisfactory. Everywhere I went, from north to south and from east to west, I found that villages and cultivation had increased. The population is larger and wealthier ; flocks and herds are more numerous ; security prevails, and general satisfaction is expressed with the present rule.' Once more, or, rather, nearly for the first time, life and property are safe. Relieved from the scourge of war and tyranny, people are everywhere resuming their old occupations. They have even recovered from the shock of finding themselves under just and settled government, and are no longer content merely to exist. New wants are being felt, and with the advance of material prosperity trade and commerce are springing up.

There are, however, considerable obstacles to the development of trade. First of all there is a lack of labour. Partly this is due to lack of population, but partly also to other causes. Slavery has left its mark, and many of the Arabs are too proud and too lazy to take part in manual labour ; in laziness, though not in pride, many of the negro tribes are fully their equals. Secondly, there is the difficulty of communications in so vast a country, and the lack of transport. Thirdly, the Soudan is very poor, and capital is wanting. Still, every year shows an improvement in these respects.

The population is steadily growing, partly by natural increase, and partly by immigration from neighbouring countries of people who had fled during the rebellion. Attempts have also been made to assist the increase by colonization. A number of old soldiers from the Soudanese battalions, who are enlisted for life, were permitted to retire, and with their wives and children were established in villages on the Nile and at Kassala. The villages were organized on a more or less military basis, with a well-known non-commissioned officer as chief. Each colonist was allotted two or three acres of good rain or pasture land, or an acre of Nile foreshore. He was given grain for sowing, besides a quantity of dhurra sufficient to support him until his crops grew. Markets were also started. Unfortunately, the experiments were generally unsuccessful. The colonies at Dongola and Berber failed altogether. It was found that the black when released from the strict discipline of the regiment was more anxious to enjoy doing nothing, after the manner of his ancestors, than to work, and if he saw a chance of living by begging or stealing he was apt to leave his cultivation alone, and go off to some town.

At Hellet Abbas, on the White Nile, it was found that if the rains were good the colonists would prepare the ground and sow the crop, but if it came to artificial irrigation and shadoof work they soon tired of this heavier labour and left the crops to wither. Happily, a much better account comes from Kassala. The Mudir of that province reported:

'The colony of blacks established at Kassala continues to thrive, and in every way justifies its existence. They have a well-laid-out village, and are eager to cultivate along the Gash, and have also a fair amount of rain crops. Labourers can nearly always be obtained from amongst them for public works, and there are some very fair masons who are permanently employed. I wish we had the means of teaching some of them carpentry and blacksmith's work as well. They have acquired a good deal of small stock.

'Those of Gedaref are not so thrifty, and are lazy; they do not cultivate so much or so well, but I hope for improvement.'

Transport and communications are indeed vastly better than they were in Egyptian days. The railway from Halfa to Khartoum makes an all-important difference. It has already given an immense stimulus to the export trade; without it there would have been quite another tale to tell of the last five years. But except in the province of Berber it does not tap any local resources. Great part of it lies in an uninhabited desert. Every train leaving Halfa has to carry with it 1,520 cubic feet of water. And there are 200 miles of river between Halfa and the outer world at Assouan, and then 700 miles more to the sea. Carriage of goods over such a mileage, with its necessary transhipments, is a long and costly business, nearly prohibitive for bulky articles. Coal, for example, is seldom less than £4 a ton at Khartoum, and often nearer £6. No one who had to build a railway in the Soudan for commercial purposes only would think of crossing the desert to

Halfa; his first thought would be to connect Khartoum with the sea-board at Suakin.

There is also the railway from Halfa to Kerma, thirty miles from Dongola. But this, too, was laid down in haste for military purposes. It is laid so badly, with such sharp curves and such steep gradients, owing to the nature of the country, that no heavy trains can run on it, and with the present rolling-stock an engine has sometimes to make three or four starts before it can master an ascent. It is worked at a loss of about £20,000 a year, and it has become a serious problem whether the money so spent could not be employed much more advantageously elsewhere, so many are the claims on the Soudan Exchequer. It would be far better to take up the whole line, and relay it from Dongola to connect with the main-line at Abu Hamed. For not only is this a much easier country, but the southern part of the province, which is the richer, and inhabited by a more industrious and hard-working population, would be opened up. But at present no money is forthcoming either for this or for the complete repair of the existing line. So poor Dongola is in the tantalizing position of having a railway, and yet not being able to take full benefit of it. It is actually suffering from a surplus of food-stuffs, and is a year behindhand in its exports.

The province is famous for its dates; not only is it a great exporter of the ordinary fruit, but it also produces a golden date, which is said to be the best in the world, better, even, than the Algerine date, so well known in Europe. It has also abundance of irrigable land as good as Upper Egypt, which it very much resembles in general climatic conditions. As its numerous and interesting antiquities show, it once supported a very large population; but now a great deal of land is lying waste, and the population, though increasing very fast, is still not more than about 100,000. With better communications its prospects are very good. When Egypt becomes overcrowded, as it must in time if the present rate of increase is maintained, Dongola will offer a fair field to Egyptian immigration. The conditions of life are so similar to those in Egypt that it cannot fail to be the most attractive part of the Soudan. Even as things are, owing to the partial advantage afforded by the railway and facilities of transport by boat-carriage on the Nile, the trade of the province is increasing. Cotton goods and luxuries like tea, sugar, coffee, and perfumeries, are the principal imports, and cereals are exported as well as dates. The people are wealthier, and anxious to buy such goods as cutlery, crockery, soap, agricultural implements, and hardware, but well-to-do traders have not as yet exploited the field, as they would do if the railway difficulty were solved.

Once Khartoum is connected with the sea by railway, the principal obstacle in the way of trade will have been removed, and, fortunately, this is no longer a mere vision of hopeful men. Practically the railway has already been begun. It will strike across the desert to the Atbara, crossing the mountains near Sinkat, and then run along it to join the main-line near where it crosses that river. If the railway does not come to Berber itself, Berber will probably travel up the Nile to meet it. Arab towns are not very difficult to shift. The whole route has been carefully surveyed, and a good deal has been spent in improving the port of Suakin. Materials are being rapidly collected. This time careful preparations are being made; there will be none of those kaleidoscopic changes of policy which were so fatal in 1885. It may confidently be expected that in two or three years' time there will be something to show very much more substantial than a tennis-court at Dover built of much-travelled material, the only result of our former exertions. For the construction of the line the Government will be its own contractor. The Soudan Railway Department has a very capable staff, and they will be able to do the work as efficiently as any outside contractor, and much more cheaply. It is hoped that as far as possible native labour may be made use of. It is a hopeful indication of a change of spirit among the Arabs that the local sheikhs have agreed to bring their tribesmen to work; the experiment is worth trying even if it fails. No such thing has ever happened in the Soudan before.

Meantime Suakin is looking eagerly forward. Its inhabitants have naturally suffered by the complete diversion of trade to the Nile Valley route. Very few of them can even afford to repair their houses, and the town shows signs of decay. Most of the people are unemployed, and labour is very cheap. There was formerly a good deal of trade with India and the Red Sea ports, but most of this has fallen away, and nearly all the Indian merchants who formerly had their headquarters there have left. But now that the desert which shuts off the Soudan from the sea is really to be bridged over, there will be a great change. The one seaport of these immense territories cannot fail to be a busy and prosperous place.

Once the new line is completed, the distance from Khartoum to the sea will be reduced by about two-thirds, to some 450 miles, with a proportionate reduction in expense of carriage, and it will then be possible to think of building other feeder lines in various parts of the Soudan. A branch line to Kassala and on to Gedaref and Gallabat along the Abyssinian frontier will tap a very rich district and open up the Abyssinian trade. Possibly in the distant future such a line may be continued southwards so as to connect with Uganda. In the recent agreement with Abyssinia, powers have been taken to build in Abyssinian territory for this purpose. But this is still a long way outside practical politics. Of far more importance for the immediate development of the country would be a light railway from Omdurman or Duem to El Obeid, or across the rich Ghezireh from Duem to Wad Medani on the Blue Nile, or, again, from near Wad Medani to Gedaref. Easy communication with the sea will render it possible to bring the necessary plant into the country at a reasonable cost, and the experience gained as to material and labour in building the Suakin line will also be invaluable.

Meanwhile the provincial Governors are doing all they can to improve the caravan routes and roads. In great part of the Soudan this is simply a question of increasing the number of wells; metalled roads are scarcely necessary as yet, outside a few towns. Wheel traffic is almost non-existent; the camel and the ass are the great public carriers. It is rather strange that the camel has not been more used for pulling wheel transport than he has. One camel can pull three or four times as much as he can carry. All the heavy machinery of the Nile Valley Gold Mining Company, working in Nubia, has been transported in this way from the river, a distance of sixty miles. Rarely a camel may be seen hauling a plough in Egypt. But there are limits to the use of camels. They cannot breed successfully south of parallel 13°, and in all the country south of this the serut fly makes it almost impossible for either them or horses to live at certain seasons of the year.

Above all, access to the sea will greatly stimulate the use made of water carriage within the country. In the Nile and its tributaries the Soudan possesses a system of natural trade-routes unequalled in Africa for internal commerce. The river traffic, though already growing, is merely in its infancy. The Government has a considerable fleet of steam and sailing vessels between Wadi Halfa and Assouan, and also on the Blue and White Niles. They are also encouraging an English company which has placed some steamers and steam-barges on both rivers. On the Blue Nile there is regular steamer communication with Rosaires, 426 miles from Khartoum, during high Nile, about six months in the year, and even in low Nile most of the river is navigable by native boats. On the White Nile, now that the sudd has been cleared, steamers and native boats can ply the whole year round up to Gondokoro. In time the Sobat may prove a very good route for trade with Abyssinia. And the Bahr el Ghazal, with its network of waters, is in this respect the most fortunate of all the provinces. Many of its waterways are still blocked by sudd, but every year the navigation is improving. A serviceable channel is now available on the Jur River as far as Wau, and large steamers will eventually be able to get up even much further than this. The other rivers will in time be opened up. In a country where at present everything has to be carried on men's heads, this will be an extraordinary benefit from every point of view, and if the province answers at all to its old reputation as one of the most fertile spots of Africa, it will do very well, in spite of mosquitoes, serut flies, malaria, guinea worm, and all its other plagues.

As for the lack of capital both for private enterprise and public works, that, like the lack of population, can only be cured by time. Overhasty development could only do harm, even if it were possible. As trade improves and agriculture develops, the people will become more wealthy, and the two will react upon each other. That large sums will be invested by private capitalists or firms is not to be expected for years to come. But even now an Englishman is erecting flour-mills at Wad Medani, and if his venture succeeds it may be followed by others. The Government is doing all in its power to encourage agriculture by small loans for the purchase of seeds, water-wheels, and cattle to work them. Perhaps some form of State-supported Agricultural Bank will be established.

The Soudan is an agricultural country; in that direction alone can real progress be made, and the progress depends mainly upon irrigation. Between the Atbara and the Blue Nile, and between the Blue Nile and the White Nile, there is, as has been pointed out, a great field for irrigation works on a large scale, but the amount of crops now grown by direct irrigation is very small. Along the Blue Nile water-wheels are numerous for the first ten miles above Khartoum. There used to be 3,000 between Khartoum and Berber, but in 1898 there were no more than seventy; and though the number has increased, it has yet nothing like reached the old level. The same is true of the rest of Berber Province and of Dongola. There is also a certain amount of irrigation on islands in the White Nile above Khartoum. When the river is falling, large mud flats appear in the centre of the stream. To these the people transport their cattle and belongings; they sow their seed in the mud, build themselves huts, and set up shadoofs to water the growing crops of wheat, barley, dhurra, and onions. They begin operations towards the end of January, and in good soil the crops are harvested by May.

What perennial irrigation can do is shown by the gardens at Khartoum, where lemons, figs, oranges, pomegranates, bananas, vines, and all kinds of vegetables, grow in profusion all the year round. But at present nearly all the crops are rain crops. The height to which water has to be raised out of the Blue Nile is too great for extended irrigation by mere lift. For the first 150 miles from Khartoum the banks are 26 to 30 feet over the summer level, and further south 33 to 39 feet, the difference between summer and flood level being about 23 feet. The soil of the Ghezireh is a rich alluvial deposit sometimes 150 feet deep. The inhabitants build small dykes across the general slope of the country so as to prevent the rain running off too quickly, and sow their seed as soon as the rain has fallen. The result as described by Sir R. Wingate a year or two ago is as follows:

'I recently rode from Wad Medani on the Blue Nile to opposite Duem on the White Nile, eighty miles across a perfectly flat plain sown almost throughout its entire length with dhurra, which was standing 6 to 8 feet high. As there is only one crop sown during the short rainy season, and as this is planted and harvested within a period of sixty to eighty days, it follows that, if a system of irrigation were possible in the Ghezireh it would become a huge granary capable of supplying, not only the whole Soudan, but other countries as well.'

But, of course, any schemes for utilizing the waters of the Nile have always to be considered in reference to the prior claims of Egypt. It is only to Egypt that the Soudan can look for the money necessary to carry out great works, and, naturally, Egypt would not allow irrigation to be developed in the Soudan unless her own needs were amply safeguarded. Whenever Egypt undertakes the great works contemplated on the Upper Nile, the Soudan will share in the benefit. That will not be for a long time to come, and meanwhile the Soudan has opportunity to develop her resources and her population so as to be able to seize the chance when it comes. There seems to be no reason, however, why works on a small scale should not be undertaken, subject always to the question of expense, to utilize in the Soudan, by means of basin irrigation on the Blue Nile or the Atbara, some of the flood-water during July, August and September, which is only an embarrassment to Egypt now that perennial irrigation is adopted so universally in that country. It is only in the preceding summer months, when the Nile is low, that Egypt is forced to watch the proceedings of her neighbour with such a jealous eye. It would be rash to prophesy exactly when irrigation works will be undertaken, or what form they will take; but it is certain that they will come, and when they do, their effect upon the Soudan will be immense. The subject is of peculiar interest to England ; the lands in question are capable of producing other crops than cereals, and, as will presently be shown, for none are they more suitable than for cotton.

Quite apart from any such speculations upon the future, the external trade of the Soudan is capable of great expansion under present conditions. An examination of the returns of imports and exports during the last years before the rebellion is a very useful guide to the capacities of the country. The figures are taken from Colonel Stewart's report.

There are no returns of imports except for the port of Suakin, but these embrace practically the whole. Of course, a certain amount came in by the Nile Valley route, just as now a certain amount comes in by Suakin instead of by the railway. The position has been practically reversed, and for purposes of comparison it is fair to take the imports formerly entering the port of Suakin and those now carried by the railway. The most important item was cotton goods, which amounted to about 25,000 to 30,000 kantars annually. Linens were about 200 to 300 kantars, and woollens 100 to 300 kantars. There was also a certain amount of silks and silk thread and sewing cotton. The Indian trade brought in a good deal of grain and tobacco. In 1881 the import of Indian rice amounted to 20,000 kantars. Petroleum (6,000 kantars), oil, zinc, copper, and iron, appear in the list, as well as flour and provisions of various kinds in large quantities, and candles, boots, and clothes. The peculiar tastes of the Soudanese in luxuries are reflected in the large imports of sandal-wood, and scents and perfumery, especially fish-scent (700 tons, or 15,400 kantars, a year). Both men and women are particularly fond of strong, greasy scents. In all native festivities and entertainments these scents are a very prominent feature, and a native marriage can be smelt a very long way off. Last, but not least, comes a very peculiar item : in one year no less than 11,000 dozen umbrellas were imported. There is certainly enough sunshine in the Soudan, and in parts of it rain also, to justify a large number of umbrellas, but it is difficult to believe that 132,000 umbrellas would find a ready sale in the Soudan to-day. Perhaps the Khalifa particularly disliked the umbrella-carrying class.

In 1901 the imports of cotton goods had already reached their former level, amounting to about 28,000 kantars. But the trade is growing rapidly : in 1902 the imports had risen to 38,000 kantars. Considering that the population is no more than a third of what it was in 1881, this is striking testimony of the good effects produced by just government and a railway. The imports from India have, of course, fallen away. Scents are far below their former figure, but the imports of flour and rice have risen ; 343 tons of flour and 43 tons of rice were carried in 1901, and 733 tons of flour and 108 tons of rice in 1902. Tobacco, oil, and petroleum are also increasing. There is also a very important article of import which reflects both the improved purchasing power of the Soudan and the increased production in Egypt ; this is sugar. The railway carried 1,700 tons in 1901, and nearly 3,000 tons in 1902. Soap, likewise of Egyptian manufacture, also appears in the returns to the extent of some 140 tons a year.

As for imports so for exports Suakin was the chief channel of trade before the rebellion. But a good deal also went by Korosko and the Nile, principally gum, senna, coffee, and ostrich feathers. An inappreciable amount, probably, also filtered through to Egypt by Assouan or Assiout. Far the biggest item of the recorded trade was gum. The export for three years was—

	Kantars.
1879	144,706
1880	135,646
1881	150,861

Next came cotton, cleaned and uncleaned, averaging about 20,000 kantars per annum. Coffee amounted to about 7,000 kantars, principally from Kassala and the Abyssinian frontier. The Soudan is a land of strange diseases, but it is also a land of medicines. Besides tamarinds, as much as 3,300 kantars of senna was exported in 1880. Ostrich feathers from Kordofan and Darfur came to about 200 kantars, and there was also about half that amount of guttapercha from the Bahr el Ghazal. Miscellaneous exports included skins and hides, mother-of-pearl from the coast, mats from Kassala, and ivory.

To-day gum-arabic is still easily first among the exports. The best kind of gum is the white gum produced from the gray-barked acacia, called hashab; there are also inferior kinds called talh, or latch, produced from the red-barked acacia. The gum is used for giving a glaze to linens and other woven fabrics, for stiffening cotton-stuffs and calicoes, and for making sweets and chewing mixtures. Kordofan is the home of the gum trade; a good deal, but chiefly the inferior kinds, comes from the forests of Sennar and Kassala, but nearly all the best white gum comes from Kordofan, especially the part round Taiara between Duem and El Obeid. Here there are limitless forests of hashab. The taking of the gum does not permanently injure the tree, and the only obstacles to its gathering are the distances to be travelled without water, and the lack of labour. Every new well means a fresh productive area, and people migrate temporarily from other parts of the Soudan for the gum harvest. Most of it is brought down by boat from Duem, some directly on camels from El Obeid to Khartoum, where it is sorted on the beach. Of this first-rate white gum the Soudan has practically a monopoly, and when the Soudan was closed to trade its price went up enormously. This naturally brought out substitutes: inferior gum from West Africa came into use, and glucose also took its place. Still, there was a good demand for it after the re-occupation of the Soudan, and in 1900 the price was still 65s. per cwt. Since then the production has gone up by leaps and bounds, but, unfortunately, this has been accompanied by a great fall in prices, at one time as low as 27s. 6d. per cwt. The figures of the export trade are:

	Kantars.
1899	41,963
1900	60,912
1901	170,781
1902	220,000

The trade has now far outstripped its former limits. To a young country, a profit of some £200,000 a year divided between the merchants and producers on the one side and the Government on the other by means of royalties and railway receipts is no mean advantage.

As for the other former exports, the trade in

ostrich feathers from Kordofan and Darfur has begun again, and there seems no reason why it should not be developed. Ostriches are farmed successfully in Egypt near Cairo, and the conditions are even more favourable for their establishment in the Soudan. Nor is it unlikely that the Soudan will be able to supply a part of her own tobacco and sugar, which now bulks so largely in the imports. In former days the sugar-cane was cultivated largely in Dongola and along the Nile in Berber Province. The fertile plains around Kassala bore crops both of sugar-cane and tobacco. The district of Fazokhl, beyond Rosaires on the Blue Nile, used to produce 1,000 kantars of tobacco per year. It is also found in Fashoda; and in the south-western part of Kordofan, where the soil is the richest in the province, both tobacco and sugar-cane grow easily wherever there is water, to say nothing of the Bahr el Ghazal. The coffee came principally from that part of the country which has since been taken over by Italy or Abyssinia, and, though it is grown in Kassala, the trade in it is not likely to come to very much. This coffee, however, which is of the Abyssinian kind, and not a first-class coffee, is still quoted in the London market at about 50s. per cwt.

But the most promising feature in the old returns is the 20,000 kantars of cotton, even though as yet the trade has not revived. Though the figure is in itself insignificant, it is a proof that the thing can be done. Cotton is indigenous in the Soudan. It grows wild in Fashoda, although the native Shilluks seem never to have taken advantage of this circumstance, preferring to go completely naked. Most of the former cotton export came from Kassala. The Khor Gash, a tributary of the Atbara, comes down in flood during July and August, and partly inundates the plain, leaving behind a rich alluvial deposit—splendid cotton soil. There was formerly a cotton factory in Kassala town. In the districts of Gallabat and Gedaref cotton is now actually being grown, and it is proving the foundation of increasing trade with Abyssinia. Abyssinian merchants eagerly buy up all that can be grown. Further south, in Sennar Province, the valleys of the Dinder and the Rahad were once very famous for cotton, which was also largely imported into Abyssinia. Here, too, the cultivation is increasing as the people settle down. The district of Tokar, near Suakin, along the Khor Baraka, produces, perhaps, the best quality of all the cotton in the Soudan. It, too, in former days produced very much more than now. While along the Nile itself, in the neighbourhood of Khartoum and in Berber and Dongola Provinces, enough cotton is grown to supply small local industries, in which a rough white cloth is woven, one of the few local manufactures in the Soudan. Beyond a doubt, not only is there a great deal of land admirably suited to grow cotton in the Soudan, but also the climatic conditions, given only water, are peculiarly favourable.

Cotton and its culture are thus no novelties to the inhabitants of the Soudan. The point is the water; it all comes back to water and irrigation. If the Soudan is to be of any real interest to the cotton-spinners of Lancashire, its export must be counted not by a few paltry tens of thousands of kantars, but by hundreds of thousands, perhaps by millions. And for that there must be irrigation works on a large scale. The soil is there in the Ghezireh and elsewhere, the climate is there, the water is there, and the irrigation works will come. But once again, there is no need for hurry. The interests of the whole Valley of the Nile have to be considered. The undertaking is too large to be gone into without the utmost care and patient deliberation.

It is eminently satisfactory that the Government is fully alive to all the possibilities. They have started an experimental farm at Shendi, where trials are being made of different sorts of cotton, of different methods of culture, and of different periods of sowing, as well as calculations of the cost of production and of carriage to the ginning factories in Egypt. Already some most interesting and important results have been obtained. It has been definitely shown that the cotton which is sown in June and July promises better, both in quality and quantity, than that sown in the autumn or in March and April. At that time the heat is not so great, and the river is rising, so that the cultivator gets his water during the most necessary time at the least cost, because with the least effort. If this is confirmed, it is extremely important, for the water will be taken at a time when the Nile is high, and when, therefore, Egypt can afford to allow it to be used without suffering in the least degree, apart altogether from new Reservoir works.

As regards quality, it appears that the cotton grown, if not so good as the very best kinds of Delta cotton, is at least as good as, or better than, the best American, both in colour and staple. It is calculated that at the present time 1 acre producing 4 kantars will produce gross receipts of 1,060 piastres, against an expenditure of 1,000 piastres, showing a profit of 60 piastres, or 12s. 6d. per acre. But when the new railway has reduced the cost of fuel for the pumps, and also the cost of carriage, the expenses will be no more than 700 piastres, showing a profit of 360 piastres per acre, or 75s. It is estimated that the new railway will reduce the cost of freight by 50 piastres per kantar, and, wherever the Soudan has its own ginning factories, the profits will, of course, be all the greater, because only the prepared product will be carried. There is at present sufficient local demand for cotton to make it generally more profitable to sell it on the spot than to carry it to Egypt, but as the production increases it will soon outstrip the local demand. Any private capitalist investing money in cotton in the Soudan would be able to buy and clear land on the river in Berber or Dongola at from £5 to £6 per acre, so that he would get a very reasonable return on his investment. He would have the further advantage that in the Soudan two of the worst cotton diseases, 'worm' and 'hog,' are unknown.

Cotton and corn are the two great foundations on which the hopes of commercial prosperity in the Soudan are founded. The Negro Soudan is still comparatively unexplored, and its resources cannot be estimated. There is, however, a chance that the Bahr el Ghazal will do great things in rubber. Rubber-trees are known to be plentiful; rubber has already been produced in small quantities, and specimens of absolutely first-class quality have been obtained. But it has yet to be shown that the best kind of creepers are numerous, and also that they can be successfully tapped without killing the plants. So far, it does not seem likely that the Soudan has any great sources of wealth underground. Iron is found and worked in small quantities in the Bahr el Ghazal, and at least two ore beds are known in Kordofan, but there is no fuel to work them, and no means of transporting the ore. As for gold and the precious metals, several prospecting licenses have been issued and search is being eagerly made, but, except the copper mines of Hofrat-en-Nahas, nothing has been discovered at present. The only known gold-bearing district, the Beni-Shangul, is now included within the territories of Abyssinia. Gold was formerly worked in this district by the Egyptians, not at a profit, and perhaps, in any case, it is no great loss to the Soudan (even if it had not already been occupied by the Abyssinians), for that gold may not be discovered in the Soudan is the earnest prayer of every official in the country. The true wealth of the Soudan, such as it is, lies in its water and its soil. A find of coal would be a very different matter, and much more valuable than gold, but, though discoveries are constantly being rumoured, coal is not yet.

No account of the commerce of the Soudan would be complete without a mention of those wonderful people the Greek traders.

It is well known how, on the day after the Battle of Omdurman, a Greek arrived in the town and opened a store with all kinds of goods brought somehow from Suakin. This man is now a prosperous and wealthy merchant, with large shops in Khartoum and Halfa, and a finger in every sort of commercial undertaking. He is no longer alone in the field. Whether it be true or not that trade follows the flag, undoubtedly the Greek trader follows the British flag. They are said to be principally Ionian Islanders, so perhaps they have a hereditary liking for it. Just as from Alexandria to Halfa every town in Egypt has its Greek traders, carrying on business as storekeepers, dealers, and innkeepers, so, from Halfa to Gondokoro, from Suakin to El Obeid, every town in the Soudan has its Greeks. They are ubiquitous; in Khartoum and Omdurman alone they number about 800. One wonders what they were all doing before the Soudan was open. Some of them, no doubt, stayed on right through the Khalifa's time.

As a vulture scents carrion from afar, so the Greek scents any possible opening for trade with the natives. The gum trade, the feather trade, the corn trade, all are in his hands. There is nothing that a native wants, however humble, from beads and kerosene-tins to silver, that he will not sell, exactly in the form required. Naturally a gambler, there is no speculation that he will not undertake, no risk he will not run. He can stand any climate, he can live in native huts, and eat native food. He may be unscrupulous in his dealings, and he has to be sharply watched by those in authority, but as a trade pioneer in a new country he is invaluable, and his enterprise contributes largely to make life possible for more exacting Europeans in desolate places. Some day, perhaps, when the Arab has mastered his methods of trade, he will find his match, but at present he holds the field.

CHAPTER XXI

TAXATION, REVENUE, AND EXPENDITURE

At the beginning of every year the Soudan Budget, with the estimates of receipts and expenditure for the year, has to be submitted to the Egyptian Council of Ministers. The total amount to be granted to the Soudan from the Egyptian revenues is then decided, and any consequent alterations in the Budget made. The Governor-General and the Financial Secretary are responsible for seeing that the total sum so granted by Egypt is not exceeded, but while this limit is observed, credits can be transferred from one head to another without reference to Cairo. Any special or unforeseen expenditure can be defrayed by Egypt by special grants sanctioned by the Egyptian Council of Ministers, and the Egyptian Minister of Finance has at all times the right of inspection, audit, and supervision of the Soudan Government accounts. Practically, of course, this means that the finances of the Soudan are subject to the control of the British Financial Secretary and the British Consul-General at Cairo, but that the Soudan Government exercises a very wide discretion as to the disposal of its revenues once realized.

The revenues of the Soudan are therefore at

present made up from three sources: first, taxation, fees, licenses, etc.; secondly, the receipts of certain earning departments—*e.g.*, railway, steamers, post, and telegraphs; and thirdly, the contribution from Egypt. Of these the third is still, it must be confessed, the most important, but the first is steadily growing. Strictly speaking, the second should not be included, for although the earnings of these departments are materially increasing, and although without them very little could be done, their annual accounts do not as yet show a profit, for various reasons that will be explained.

The principles of taxation are not new. Many of the taxes are the same in kind as those which were formerly in force not only under Egyptian rule, but also under the Khalifa's. It was wisely decided that no innovations should be introduced based on Western notions, unless they were unavoidable. The taxes are of a kind to which the people have long been accustomed, but, of course, great care is taken that they shall be fixed at a moderate rate, and that no one shall pay more than is actually required by law. It was the uncertainty of the amount to be paid, the illegal imposts levied by the collector on his own account, and the dates at which they were collected, that made the old taxes so often cruel and ruinous. The uncertainty of the amount to be paid had, however, an attraction for the Oriental. It varied with the circumstances of the year. The Government wished to get all they could, and in a good year they exacted a most excessive amount; but in a bad year they took little or nothing at all. According to the Western system a fixed moderate amount has to be paid every year, and a whole or partial failure of the crop is not considered any excuse. The latter system is, of course, far the most just and economically sound, but it lacks the elasticity of the other. It is the constant endeavour of European administrators in Oriental countries to endeavour to combine the advantages of both systems. In the Soudan the system of petitions direct to the Governor-General is a means to secure this result. When adequate cause is shown, he is able to remit taxation wholly or partly, and there have been several cases in which he has done so, notably in Berber and Dongola Provinces in 1900, when the low Nile caused a great deal of land to be thrown out of cultivation. Of course, such remissions are inclined to play havoc with a Budget framed on careful estimates. It cannot be helped, and at any rate it is only one of many difficulties which have to be encountered in balancing a Soudan Budget. Unforeseen expenditure often becomes necessary, but happily during the last two or three years the receipts have been increasing unexpectedly also.

Although the taxes remain the same in principle, a comparison with the former state of things shows the difference in the methods of assessment and collection to be so great as to amount to a revolution. Then, as now, the Government in the Soudan was, according to the general rule in the East, the admitted owner of the soil, and the cultivator had to pay a tax amounting to one-tenth of the produce for its use. The actual sum to be paid included all other taxes, as house-tax, animal-tax, and so on. Each district was rated at so much in the Budget; this sum was divided among the villages and communes until the individual cultivator was reached. The nomad and other tribes which did not cultivate the soil were assessed at an annual tribute, according to their wealth in camels, cattle, and horses. In Berber and Dongola the tax was not on the produce of the land, but on the instruments of irrigation. It varied according to the instrument and according to the quality of land. One sakieh was calculated to be capable of watering 8 acres, and if more land was cultivated an increase was made in the tax. It has already been mentioned how these water-wheels were often taxed in a sum far beyond their possible earning capacity. It was the same with the assessment of the districts and the tribute of the tribes. The official estimates were seldom or never realized as far as the public exchequer was concerned (although the bullying tax-collectors took much more for themselves). Arrears accumulated, in spite of every effort, and became a direct incentive to discontent and revolt. In the year 1879 the deficiencies of the amounts actually received below those estimated ranged in the different provinces from 10 to 60 per cent. The figures of the Budget were, in fact, pure fancy, especially when the country was disturbed. But the thing was even worse than it looked. Districts and tribes, though they might not pay all that was asked, paid a great deal more than they could afford, and consequently, as the money had to be found somehow, they found it by indulging in the only prosperous trade—that of slavery. In other words, the Soudan lived largely by expending its capital.

Owing to the alteration in the boundaries of the Soudan, it is somewhat difficult to calculate what the amount raised in taxation from the present territories amounted to. As nearly as can be made out, it was about £360,000 a year nominally, with a nominal deficit on the whole administration in the same provinces of £70,000 a year. Absolutely nothing was spent in the development of the country ; salaries, rations, and clothes for the army of occupation made up almost the whole of the expenditure.

As compared to this, the actual sums raised in the Soudan by taxation during the last four years have been as follows :

	£E.
1900	95,000
1901	155,000
1902	190,000
1903	200,000*

But the conditions are so different that nothing can really be deduced from this comparison. The figures show clearly that the country is advancing steadily. The burden is lightly borne. It would be very easy to exact far more than this without any positive discontent, but it would be a most unwise policy to do so, for it would be at the expense of its future progress.

The taxes are no longer assessed by the district. The tribute from the nomad tribes is the only exception to the rule that the Government deals directly with the individual taxpayer. They are no longer collected by irregular and irresponsible bullies; they are payable at the most convenient seasons, and they are moderate in amount. So great a difference in detail is almost a difference in prin-

* Estimated. £E1 = £1 0s. 6d.

ciple; but the names are the same. Of the whole collected the receipts from land-tax and ushur are about a quarter—*i.e.*, about £50,000 a year. The land-tax proper is no longer levied according to the instruments of irrigation, but, as in Egypt, according to the acreage and situation of the irrigated fields. Land irrigable by wells and foreshore land irrigable by flood pays 20 piastres (4s.) per acre; land on the mainland irrigable by water-wheels and shadoofs, 30 to 40 piastres; and land on islands similarly irrigable, 50 to 60 piastres. All other lands—that is to say, lands which depend for their water-supply not on irrigation, but on rainfall—are taxed at the rate of 10 per cent. *ad valorem* on their produce. This tithe is called ushur. At first it was paid almost entirely in kind, but it is now collected as far as possible in money, and, as the money in circulation increases, payment in kind—a most inconvenient form of payment—will be altogether abolished. The date-tax—2 piastres (5d.) on every date-palm that has begun to bear fruit—is akin to the land-tax. It produces some £15,000 a year.

The land-tax in some form will be the staple of the future revenues of the Soudan. Every year as population increases and more land comes under cultivation it will form a larger proportion of the total receipts. In many of the remote parts of the country, owing to the great distances and the small number of officials, it is hardly yet in working order, but with improved organization all may be expected to bear their just share of taxation, and the receipts will benefit accordingly. Any large scheme of irrigation and cotton culture will, of course, send up this branch of revenue by leaps and bounds. Second in importance to the land-tax, and for the time being exceeding it, come the royalties charged on exports of gum, ostrich-feathers, ivory, indiarubber, and a few other articles, at the rate of 20 per cent. by weight. The extraordinary growth of the gum-trade is mainly responsible for this item, which makes up another quarter of the taxation revenue. Ivory receipts can hardly be expected to come to very much, but ostrich-feathers and rubber will probably show a marked increase in future years.

Other minor receipts were estimated in the Budget for 1902 as follows:

	£E.
Animal-tax	9,500
Tribute from tribes	8,000
Ferries and fisheries	4,000
Sale of salt	1,530
Stamped paper	1,200
Customs	5,400
Slaughtering dues and market fees ...	5,200
Road-tax	2,000
Rent of Government lands	1,300
Boat-tax	1,000

All these taxes are familiar in the Soudan. Some of them, especially those small in their results—and there are a good number more of less importance than those named—will probably be abolished as soon as the country is able to stand the immediate loss. But as yet even the most objectionable in principle do not act as restrictions to trade. They only distribute the burden of a very light taxation, so as to make all classes contribute something, and the Government is still so poor that even a few hundred pounds are of very great importance.

The Customs are set down as producing only

£E5,400, and this demands a word of explanation. £4,500 of this is taken at the port of Suakin, the remainder on the land frontier in Kassala and Sennar. The Customs on the land side, small as they are, show a steady increase; those at Suakin have been constantly decreasing, as trade is more and more diverted to the Nile Valley route. But all Customs duties on goods coming to the Soudan on this side are levied at the Egyptian ports, and retained by Egypt. The duties are 20 per cent. per kilo on tobacco, and 8 per cent. *ad valorem* on all other goods. In the year 1903 it is estimated that the sum thus retained by Egypt will amount to £E70,000, but this has not been taken into account in calculating the Soudan revenues.

As for the profit-earning departments, their profits are as yet invisible—invisible, that is, as far as their accounts are concerned, though visible enough in the increasing prosperity of the country. Still, the actual takings of the railway and the post and telegraphs mark progress. They stand for three years:

	Railways.	*Post and Telegraphs.*
	£E.	£E.
1900	38,412	7,900
1901	75,808	9,000
1902	85,000*	11,000*

But these receipts are far from being any criterion of the actual amount of work done. They exclude all that was done on Government account. Up till 1903 all passengers, goods, and messages on behalf of Government were carried free of charge. But it was found that this system tended to extravagance.

* Estimated.

A department, for example, wishing to buy dhurra for Khartoum, was apt to buy it at Dongola, possibly at a cheaper rate, and have it brought by rail for nothing, rather than buy it locally and disburse something for the cost of local camel or mule transport. This was good business for the department, which had only a certain credit allotted to it, but waste from the point of view of the railway, by which the cost of transport was borne. Now each department is charged in the books of the railway or post-office for all services actually rendered. The change is, of course, only one of book-keeping, but it is a good instance of the way in which good book-keeping works towards economy. With this alteration the working of the profit-earning departments makes a much better appearance. The estimates for 1903 are:

	Receipts. £E.	*Expenditure.* £E.
Railways	143,970	143,777
Post and telegraphs ...	24,428	34,800
Steamers and boats ...	69,028	86,223

The maintenance of the army in the Soudan is the item most affected by this alteration. The sum to be paid to the Egyptian War Department is set down at £E193,658 for 1903 as against £E122,548 for 1902.

The Steamers and Boats Department has been too recently organized for an opinion to be formed upon its working, but the Railways are soundly and economically managed. The increase of traffic has made it possible to reduce the ratio of working expenses, though the high price of fuel is still a great obstacle. Besides the ordinary expenditure there has also been a good deal of capital expenditure, which was very necessary considering the haste with which the line was laid down as a purely military railway. These sums are: In 1899, a special credit of £E390,000 for the completion of the line from the Atbara to Khartoum; in 1900, a special credit of £E15,000 for culverts and bridges on the same portion of the line, and a loan of £E55,000 for general development and purchase of rolling-stock; and in 1902-1903, a loan of £E528,000 (spread over five years) for the same purposes; and a special advance of £10,000 for the survey of the proposed Suakin-Berber line. All these sums, together with an advance of £E31,000 for the improvement of the harbour at Suakin, have been found by Egypt. The Soudan pays 2½ per cent. on the loans.

The nature of the country makes it inevitable that the postal service should show a loss for some time; but the telegraph service would show a profit over actual working expenses, but for the fact that so much is expended every year on extension, and this is credited to ordinary expenditure. £E18,500 was, however, borrowed from Egypt in 1900 to meet a special difficulty. The wooden poles were sometimes devoured by white ants, and were also liable to rot. After various experiments it was found best to bolt the poles on to steel bases, and it was to meet this emergency that the loan was contracted. The new plan has been found to answer admirably, and it has also facilitated telegraph extension, because the shorter poles (12 ft. 6 in. instead of 18 ft.) make a much more convenient load for a camel. Besides the wires along the railway, there is now telegraphic communication from Berber to

Suakin, Suakin by Tokar to Kassala, Kassala to Gedaref, and Gedaref to Gallabat. A line from Gedaref connects at Messalamia with a line down the Blue Nile from Khartoum by Wad Medani to Sennar, which then crosses over to Goz Abu Goma on the White Nile, and is continued to Fashoda. Another line runs from Khartoum to Duem, on the White Nile, and thence to El Obeid. In time the telegraph will be continued south to Uganda, and whenever this takes place the telegraph tariff convention arranged by Mr. Cecil Rhodes for through communication between Alexandria and Cape Town will come into force.

If the gross takings of the railways, post and telegraphs, and boats, are included in the revenues of the Soudan, the Budget wears a more imposing aspect than if only the net expenditure of each of these departments is included. The estimates for 1903 stand thus :

		£E.	£E.
Expenditure—Civil	...	624,226	
Military	...	193,658	
			817,884
Receipts	... ...		428,163
Deficit	... ...		389,721

By starving the administration it would be possible to make this deficit a great deal less, and, on the other hand, it would be very easy to make it a great deal more. A country which has been going steadily, even rapidly, backwards for so many years affords unlimited opportunity for capital expenditure. Indeed, a large part of its ordinary expenditure is really capital, so far as it is incurred for permanent buildings, railways, telegraphs, and all the other machinery, not only of government, but of elementary civilization, which were all entirely wanting. The item of public works bulks very large in the civil expenditure. It is inevitable that the expenditure should increase with the development of the country, but it is a satisfactory symptom that it is not increasing so fast in proportion as the revenue. The civil expenditure, too, goes up, while the military expenditure goes down. In the following table only the net expenditure on railways, etc., is included:

EXPENDITURE.

	Civil.	*Military (including Gunboats, etc.)*	*Total.*
	£E.	£E.	£E.
1899 ...	230,000	281,000	511,000
1900 ...	271,000	282,000	553,000
1901 ...	330,000	222,000	552,000
1902 ...	350,000	193,000	543,000
1903 ...	380,000	193,000	573,000

The richer provinces in the Soudan might prosper very much faster if they were allowed to devote the whole of their own revenues to the development of their own resources. But, of course, the whole receipts have to go into the common purse, and then be doled out again in the interests of the country as a whole. Naturally, there is a tremendous fight every year between the Mudirs, who are responsible for the provincial budgets, and the central Treasury. Two provinces stand out prominently in the matter of profits—Kordofan and Dongola. Dongola had pride of place at first, but then Kordofan passed in a stride, through the revival of the gum trade. But Dongola is still a good second, in spite of her railway difficulties, through the date-tax and the land-tax. Two-thirds of the whole amount raised by this latter tax is supplied by her. Halfa Province, or, rather, District, also shows a good profit by means of the land and date taxes. The Ghezireh Province, the kernel of the Soudan, makes a favourable appearance. Sennar just pays its way. The others all show a deficit. Khartoum, of course, stands in a different category, as also Suakin. But Berber and Kassala may soon be expected to redeem themselves. Fashoda is as yet very undeveloped, and naturally the Bahr el Ghazal, only just occupied, comes well at the bottom of the list. That this order of financial merit will be maintained unchanged is most unlikely.

It will be noticed that the estimated deficit for 1903 was some £E390,000, and this is Egypt's contribution towards the current expenses of the year. The same sum was contributed in 1902, and it has been agreed that this subsidy shall be continued at the same figure for some years to come. The approximate amounts so expended in former years were: 1899, £E385,000; 1900, £E417,000; and the same in 1901. On the basis of this contribution, which is called 'Insuffisance,' the estimates for the Soudan are drawn up. Considering all the difficulties in the way of accurate calculations, it speaks well for the Soudan officials that their expectations have been so nearly realized. In 1899 and 1900, when the accounts were finally made up, there were found to be deficits of £E23,000 and £E40,000. But in 1901 the finances of the country took a great turn for the better, and, although expenditure exceeded the estimate, this excess was so much overbalanced by unforeseen increase of receipts that sufficient surpluses were realized in 1901 and 1902 to make good the deficiencies of the two previous years. During the years 1899–1903 Egypt has therefore expended close upon £2,000,000 in these annual grants; she has also paid over, partly as special credits and partly as loans, another £1,000,000 for capital expenditure on railways and telegraphs, as already mentioned. There are deductions to be made from this sum of £3,000,000 in calculating the actual cost imposed on Egypt by the Soudan; but as these are the actual amounts on which the Soudan Budget is calculated, they must be set down here.

It sounds, perhaps, a strange thing to say that the Soudan finances are in a sound condition when such sums have to be contributed by Egypt, but such is certainly the fact. The great feature is the recuperative power of the country, so markedly displayed since 1901, and there can be no doubt that if, with Egypt's help, the two principles of low taxation and irrigation can be steadily applied, they will produce, if not as great, yet a similar result to that which they have produced in Egypt itself.

CHAPTER XXII

THE COST OF THE SOUDAN TO EGYPT

THE Nile Valley presents some peculiar examples of political organization. At one end of it is Uganda, a British Protectorate under the administration of the Foreign Office. At the other is Egypt, nominally an independent despotism, tempered by international Boards of Control in several departments, notably finance, and paying tribute to a suzerain Power — Turkey — but in the military occupation of England. Between comes the Soudan, where, except at Suakin, the British and Egyptian flags fly side by side. It is ruled by a Governor-General, joint representative of the King and the Khedive, whose acts have to be approved by the British Consul-General at Cairo, and it is jointly occupied by the troops of both Powers. If it is difficult to decide upon the exact position of Egypt, what shall be said of the Soudan? International jurists have in it a fair field of problems on which to exercise their wits, not to mention the peculiar status of the Lado enclave. But whatever imaginary difficulties may be conceived, one thing is certain: the hoisting of the British flag leaves no doubt of the fact that the Soudan is not in any way a part of the Turkish Empire. The difficulties and complications which have been caused in Egypt by the Capitulations and the consequences derived from them definitely cease with the boundaries of Egypt. It is easy to see that those who framed the Convention of 1899 were thoroughly determined on this. Article VIII. lays down that the mixed tribunals shall not extend to the Soudan, except Suakin, and, to further preclude any chance of international trouble, it is expressly stated that no Consuls or other foreign representatives shall be permitted in the Soudan without His Britannic Majesty's consent, and that no special privileges shall be accorded to the subjects of any one or more Powers to trade, reside, or hold property, within the limits of the Soudan.

Otherwise, there was no reason why the Soudan should not have been placed in exactly the same position as Egypt, whatever that may be. In every other respect it is, in fact, in the same position as though the Egyptian flag flew alone, especially in the matter of cost. When the British taxpayer is looking at a map of the world in order to get some satisfaction for his Imperial expenditure, and casts his eyes over Africa, he doubtless comforts himself with the reflection that in the Soudan England governs, but Egypt pays, and wishes that other portions of the Empire were managed on similar principles. England governs, and Egypt pays, but the division of labour is not unfair. If the Egyptian finds it hard to realize the meaning of a veiled Protectorate, the more unsophisticated Soudanese would find it totally impossible. The flag is a visible symbol which appeals to him directly, and contributes largely to the maintenance of peace. Many a tribe submits contentedly to British dominion, which would indignantly scout the idea of submission to Egypt alone. As a matter of fact, the British Government does contribute something in money, for it bears the cost of the British battalion at Khartoum and its barracks. Still, it is to the Egyptian Treasury, and not the British, that the Soudan has to look, and it is worth while to estimate rather closely the sums which Egypt has to find, and the advantages which it gains in return.

The cost of the Soudan Campaign from the opening of the Dongola Campaign in 1896 to February, 1899, was £2,345,345, made up as follows :

	£
Railways	1,181,372
Telegraphs	21,825
Gunboats	154,934
Military expenditure	996,223
Total	£2,345,345

Since then, up to the end of 1903, the ordinary expenditure of Egypt in the Soudan has been, as was shown, about £2,000,000, and the capital expenditure about another £1,000,000. In other words, the Soudan has cost Egypt nominally about £600,000 a year for the last five years.

But as a matter of fact the real cost during the five years has been a good deal less than this. Various deductions ought to be made. In the first place, Egypt takes all the Custom duties on goods going through to the Soudan. These came to about £60,000 in 1902, and were estimated at about £70,000 for 1903, an estimate very likely to be below the mark. It is safe to reckon that during the five years they amounted to nearly £200,000. Then, the expenses of the Egyptian army of occupation in the Soudan are charged to the Soudan Government. During the first two years these amounted to about £300,000 per annum. Taking an average of £200,000, we get a sum of £1,000,000 under this head. If the Soudan had not been reconquered, Egypt would have had to maintain nearly as large an army as at present, and although the cost of maintenance is naturally larger in the Soudan, it would not be unfair to deduct another £800,000 as money that must have been expended in any case. Of the capital expenditure, £600,000 was in loans for railway and telegraph development. Although the prospect of repayment of these loans is somewhat remote, they ought not to be written off as pure out-of-pocket expenses, seeing that the Soudan Government pays 2½ per cent. per annum upon them. Two and a half per cent. on £600,000 is equal to 5 per cent. on £300,000. An investment with this return would be more advantageous to Egypt than extinguishing an equal portion of her debt, and it is therefore reasonable to deduct that sum from the total expenditure. All these allowances added together make up a sum of £1,300,000, bringing the average annual cost to Egypt during the five years down to £340,000; while in addition to all this the Egyptian revenues have also directly benefited through the Railways and Post-office by the extra passengers, goods, letters, telegrams, and money orders, passing to and fro.

Still, when all deductions have been made, it cannot be denied, especially as the Soudan still calls loudly for more capital expenditure, that the reoccupation of that country, from a strictly financial point of view, though in one aspect philanthropic, is not as yet philanthropy at 5 per cent. But there are certain solid advantages which, though their value to Egypt is difficult to calculate in terms of money, are worth to her many times over the actual sum which the Soudan costs her.

First of all comes the fact, so often insisted upon, that the whole of the upper waters of the Nile are now in the secure possession of those who are responsible for her welfare. This supreme and vital necessity overshadows all others, and would by itself have forced her to undertake almost any sacrifice. In fact, as all the water in Egypt comes through the Soudan, her contribution may be looked upon as a water-rate calculated at an exceedingly low figure. Secondly, she is relieved from all fear of foreign invasion. The frontier is once more at rest, and no longer troubled even by raids. There is a vast difference between a peaceful and comparatively prosperous neighbour and a horde of furious barbarians hammering at her gates.

Further, the pacification of the Soudan enabled the burden of conscription to be diminished. The army was reduced by 5,500 men, and the period of service was reduced from fifteen years to ten, of which five have to be spent with the colours, and five in the reserve or police. Considering how much the Egyptian fellah dislikes military service, those who are affected by this change probably regard it as the greatest benefit of all.

A new field has also been opened for Egyptian trade and the employment of Egyptians. A constantly expanding market at her door for sugar and other goods is no small advantage. A good deal of the money spent in the Soudan, though lost to the Egyptian Treasury, is not lost to Egypt, for it takes the form of salaries for numbers of Egyptians in the Government service, and the money orders passing from the Soudan to Egypt show that at least a portion of it returns immediately. Moreover, the Soudan is gradually losing its old traditional terrors, and more and more Egyptians, though as yet in small numbers, are returning to settle there when their period of service is over.

Lastly, the good name of Egypt has been restored; one of the evil pages in her history has been finally turned. The country which she once ruined by her misgovernment and oppression and by her greedy haste to share in the profits of the slave-trade, and then abandoned to barbarism, has been rescued, and set moving once more on the paths of civilization and good government. It is right that in the days of her prosperity she should do something to assist her less fortunate neighbour.

It has been said that for the present Egypt will continue her annual grant to the Soudan of £E390,000 a year. But she will also be called upon to find some large additional sums. The Soudan requires capital, but has no credit of her own on which to borrow. The British taxpayer ought not to be called upon, even if he had not borne his share already, since it is not his interests, but those of Egypt, which are primarily concerned. But Egypt herself is not in a position, owing to international complications, to contract new loans, nor is it at all desirable to impose additional taxation for the purpose. Private enterprise, even if it was possible to employ it, would in the end be too expensive. But the problem is not insoluble, though difficult. Lord Cromer in his last report gives a most admirable summary of the position :

'I hope and believe that, although the difficulties are considerable, they will not prove insurmountable. A hopeful feature of the future is to be found in the fact, on which I have dwelt at some length in my Egyptian Report, that the programme of fiscal reform in Egypt is now completed. It cannot be doubted that the people of Egypt are now very lightly taxed. Strongly as I should object to any increase of Egyptian taxation for Soudanese purposes, I can see no objection whatever to maintaining such taxes as at present exist, partly with a view to providing the capital necessary for the improvement of the Soudan. Indeed, far from there being any objection, I believe the adoption of such a course to be strictly in accordance with Egyptian interests; for, until capital is spent, the Egyptian Treasury cannot hope that any considerable reduction in the present Soudan deficit will be possible. I am, of course, aware that the purely Egyptian requirements, such as improved justice and police, to which allusion is made in my Egyptian Report, must, in this connection, take precedence of the necessities of the Soudan, great though these latter be. I am, however, not without hope that, if due care and deliberation be exercised, if the projects on which capital is spent be chosen after a thorough examination of their merits and practicability, and if everything in the nature of undue haste and precipitation be avoided, money in fairly adequate quantities may eventually be found both for the improvement of the Egyptian administrative services and for the development of the Soudan.

'There can be no question as to the direction in which capital expenditure is most required. As I have said in my Egyptian Report, the construction of the Suakin-Berber Railway is absolutely essential to the well-being of the Soudan. I need only add that all the testimony which

I received during my recent visit to the Soudan strongly confirmed me in the opinion which I had previously held on this subject.'

Such words coming from Lord Cromer are full of hope and encouragement for the administrators of the Soudan. The man who found means to overcome the financial difficulties of the Reservoir works at Assouan is more than likely to surmount those of the Suakin-Berber Railway.

In the long-run Egypt herself will benefit as well as the Soudan. Of course, most of the trade now passing through Egypt will return to its natural channel by Suakin and the Red Sea. The Customs now taken at Alexandria will go directly to the Soudan, but as soon as this happens a corresponding reduction can be made in the Egyptian contribution. Nor will purely Egyptian trade with the Soudan suffer. The Nile Valley route will remain, but it will be cheaper for goods from Lower Egypt to travel viâ Suez and Suakin. The import as well as the export trade of the Soudan will be vastly encouraged, and every step forward in prosperity will make her a better market for the goods of Egypt as well as those of other countries. Once the railway is made, but not till then, there is a possibility of the revenues of the Soudan improving sufficiently to make the country self-supporting, and able to dispense entirely with any annual grant from Egypt.

It is calculated that the construction of the new railway will cost £2,500,000. Taking this as a basis, and assuming that Egypt was able to make an arrangement under which the money should be repaid by annual instalments over a period of ten years, with interest at 5 per cent., it would involve an average annual addition to her expenditure of £318,750, or a total cost of ten times that sum. If the period was twenty years, the average annual cost would be £185,625, or, say, £200,000. It is rash for an outsider to speculate on such subjects, and the figures are merely given as a rough illustration; but it seems certain that Egypt could easily bear any such burden. Nor does it appear a sanguine forecast to estimate that within ten years of the completion of the railway the revenues of the Soudan will have so greatly benefited, both by the direct cheapening of supplies, fuel, and other material, and by the development of trade generally, that at least a saving of £200,000 a year will accrue to Egypt, even if she still finds it prudent to contribute something.

What future capital Egypt will have to find must be uncertain. The whole situation will be changed by the advent of the railway. But if all the signs of the times can be trusted, whatever her expenditure may be, she will have no reason to repent of it.

NOTE.—Since the above was in print, new light has been thrown on the subject by a passage in the Note on the Budget for 1904 by Sir Eldon Gorst, Financial Adviser to the Khedive:

'The Council of Ministers has authorized the Ministry of Finance to advance out of the Special Reserve Fund the amount required for the construction of a railway to connect the Valley of the Upper Nile near Berber with the Red Sea at Suakin. The preliminary survey of the proposed line has been completed, and an estimate of its cost prepared. The total sum required, which amounts to about £E1,770,000, will be spread over a period of from three to four years, so that there should be no difficulty in meeting the charge out of the annual increment of the Special Reserve Fund, but it will be obviously undesirable to sanction any further large grants out of the fund during this period. The construction of the railway will be put in hand without delay, and if no unforeseen contingency occurs, it may be hoped that it will be available for traffic in about three years' time.'

The Special Reserve Fund is made up of the free balance remaining at the disposal of the Egyptian Government when all other claims on their receipts have been settled.

CHAPTER XXIII

CONCLUSION

SOME time ago a small detachment of Egyptian troops was passing through a village in a remote and primitive part of Kordofan. A soldier of the party, going to draw water at the well, there met, like Jacob, one of the daughters of the people, who helped him in his task. Out of gratitude for her assistance or admiration for her charms, he gave her one of the few things he had to offer, a large red cotton handkerchief. Attired in it, the damsel excited the admiration and envy of all her fellows, and from this chance seed arose a demand for Manchester cotton goods. But as the desired articles could not be purchased for nothing, the supply of charitable soldiers being limited, the inhabitants had to apply themselves to the collection of gum in order to be able to satisfy their wants. And thus the exports improved as well as the imports.

In different forms the same process is going on all over the Soudan, and the sternest admirer of Arcadian simplicity could hardly deny that the people are happier and better because of it. The country is not such an El Dorado that they are likely to be corrupted by excessive wealth. Their awakening needs can only be satisfied by means of habits of industry.

Thanks to good government they are able on the one hand securely to enjoy the fruits of their labour, and on the other hand, they are prevented from taking any short-cuts to fortune by raiding their neighbours and selling them into slavery.

If it were necessary to justify British interference in the Soudan, the fact of the repression of slavery would alone be sufficient. If the slave-trade does still exist, it does so only precariously in a few remote holes and corners. The markets of Khartoum and Omdurman and other centres are greater and more various than they have ever been before, but this one traffic, once their principal feature, is absent and gone for ever. No change could be greater or eventually more beneficial in its effects on the whole life and character alike of the slave-owners, the slave-dealers, and those unhappy tribes who involuntarily furnished their material.

The establishment of personal liberty and of security for life and property, elsewhere the commonplaces of government, is not always the result of civilization in Africa. They are new, at any rate, in the Soudan. But it would ill become the English rulers of the Soudan to take any credit to themselves if their five years' record had no more to show than this. They must not be content to be judged by comparison with any former rulers. After the Khalifa's tyranny, almost any government would seem mild and beneficent. It is not enough merely to have stanched the wounds by which the patient was bleeding to death; the foundations must be laid for his complete restoration to health and vigour. Judged by this test, however, they need not fear the verdict.

The chief difficulty is the universal ignorance and

superstition that prevail. But these ancient fortresses are being directly assaulted every day by the extension of a sound system of education. The strong administration of equal justice, and the increasing growth of trade and commerce, work powerfully in the same direction. All three agents are stimulated by the improvement of communications by river, road, and railway. There is hope in the fact that, though the people start from a very low level, their past has not been one of complete barbarism. Although Egyptian and dervish rule repressed all habits of industry and culture, the people of the country which was once the kingdom of Sennar had considerable repute as weavers, goldsmiths, curriers, and potters. They still practise a native form of inoculation for small-pox. And, to take their traditions still further back, it is said that in some places, in the elaborate dressing of their hair and the use of ornamental sandals on great occasions, they reproduce the fashions of ancient Egypt. Ed Damer, near the junction of the Atbara and the Nile, now a paltry village, had once a university of its own no less than Sennar.

Time, of course, must be allowed for these influences to work. Not in five years and not in one generation can the evils deep-rooted in the past be swept away. Though all the reforms have been introduced most cautiously, and with great regard for the prejudices and traditions of the people, it is still sometimes necessary to teach a sharper lesson, and to clearly advertise the fact that if the Government is benign and merciful it is not because it is weak. When such occasions do arise, as in the case of the last in the succession of Mahdis, suppressed a few months ago in Southern Kordofan, there is no fumbling and delay. The blow is struck at once and decisively. There is a wonderful difference between a prophet with a divine mission to regenerate Islam by force of arms slaughtering a rabble of ill-led Egyptian peasants, and an impostor unable to free himself from prison, or dangling from a gallows in El Obeid after being captured in the field and fairly tried with all the forms of law.

Such incidents, when properly handled, are only ripples on the general calm. There have been rumours from time to time of troubles likely to arise from the action of the Sheikh-es-Senussi. There were internal troubles in the kingdom of Wadai, which lies to the west of Darfur, and it was thought that these might force him to abandon his peaceful attitude, for his influence is predominant in that country. It would undoubtedly be dangerous if he were to proclaim a holy war, for he has many adherents in the Soudan as well as in Egypt. But until now his sect has been founded on rules of conduct, and he and his predecessors have steadily refused to adopt a militant attitude. The Mahdi was a good deal afraid of him, and nominated him as one of his Khalifas, but he refused to accept the position. It is now announced that the French have occupied Wadai; we have every reason to welcome them as neighbours. Wadai was one of the few remaining strongholds of the slave-trade, and its suppression there will facilitate the operations of the Slavery Department in Western Kordofan. And should any trouble arise from the followers of the Sheikh-es-Senussi or any other sect in that quarter, it will be a great advantage to have the co-operation of the French, who would have the same interest as ourselves in speedily putting it down. On the other side, now that the frontier is settled, there is not likely to be trouble with the Abyssinians, and the Italians in Eritrea and Massowah are the best of neighbours.

It may some day be necessary to interfere in Darfur, though not during the reign of the present Sultan, Ali Dinar. Ali Dinar has seen something of the world; he is a vigorous ruler, and he understands very plainly that his interests lie in keeping on good terms with the Soudan Government. But some day Ali Dinar will be gathered to his fathers, and possibly some Rehoboam will succeed him, who will cause internal dissensions, or someone afflicted with Napoleonic ideas, who will have to be chastised. Sooner or later, something of the kind is bound to happen, but it is to be hoped that a country so remote from the Nile will continue under its native rulers as long as possible.

What the future political status of the Soudan will be depends upon the course of events in Egypt. As long as Lord Cromer is there—and long may it be—everyone may feel easy. But when some lesser man has to stand as buffer between Egypt and the Foreign Office, it may make a great difference whether Egypt takes her place as definitely a part of the British Empire or not. Every year the number of British civilians in the Soudan service is growing. Great as have been the services of the officers of the Egyptian army, their employment has this disadvantage, that they are liable to be called away just when they are most required, and when they have thoroughly learned the business of administration. It has been good for the Soudan to have the choice of some of the most capable officers in the British Army for her civil work, and a spell of civil administration is excellent training for the soldier. But a glance at the names of the governors and inspectors of provinces year by year shows what a number of changes there are under the system. Some of them will probably throw in their lot definitely with the Soudan; but this can only be counted upon in exceptional cases. As administration becomes more complicated, it will be less possible for the soldier to take up the reins as he can now. Specially trained men will be more and more necessary. But at present the Soudan civil service is a water-tight compartment. Its members are in a doubtful sort of position. They are not in the Egyptian service, nor in the British Colonial Service, nor are they definitely under the Foreign Office. Up till now there has been no difficulty in recruiting the small number of capable civilians required. But as the number increases there may be some difficulty in obtaining sufficient candidates of the right stamp. Unless the country becomes very much more wealthy, it has no great prizes to offer. But if the whole Nile Valley were in practice regarded as one country, which it really is, and all under one head, with one combined civil service, there would be much more scope for able men, and each part of it would benefit by the possibilities of interchange.

It would be a great piece of organization; but, with Indian experience before us, there is no reason why it should not be a success. Direct administration ought not to be in the hands of the Foreign Office, which has plenty of diplomatic business of its own to look after. In one department, and that a very important one, the whole of Nileland will, indeed, be so organized in practice. When the great schemes for the final binding of the Nile are put in hand, they must be all under one control, and that control will be exercised from Cairo. It would be intolerable and impossible that every time it was thought desirable to open the Reservoir gates, say of Lake Albert, there should have to be negotiations between district governments or departments. The water of the Nile has made its powerful political influence felt throughout the Soudan; in time it will play its part in Uganda.

But such semi-continental speculations are unprofitable except for those who have some power to realize them. For the present the horizon of Soudan politics is bounded by the railway. From its completion all immediate plans of progress must take their start. It would be foolish to take too sanguine a view. Railway communication between Khartoum and the Red Sea will not smooth away all the difficulties of administration. There will still be years of low Nile and scanty rainfall. There will still be an abominable climate during a large portion of the year. Mosquitoes, sand-flies, and other insects will still ply their uncomfortable occupations. Dhurra will still suffer from curious and unique parasites. Cattle will be subject to murrain. Perhaps cotton, free from 'worm' and 'hog,' will discover some new foe to combat. Locusts and white ants will still occasionally devour what they ought not, and possibly develop fresh objectionable tastes. New Mahdis, unconvinced by the doom of their predecessors, may still require to be put down from time to time. Economic laws cannot be expected to work with the same certainty and freedom from cross-currents as they do in Egypt. In other words, the Soudan is a tropical country, subject to surprises: it will have its ups and downs like other places. These and similar difficulties it would have to face, railway or no railway. But beyond question, once the line is open, it will be in a far better position to grapple with them all, and to reap the benefits of the sound foundations which are now being solidly and patiently laid by the handful of Englishmen to whose charge its destinies have been entrusted by Fate. And whatever else of good or evil the future may have in store, the Soudan has one treasure which makes it certain that it will never again be allowed to lapse from the pale of civilization, and that is not the gold which attracted the attention of its former invaders, with such disastrous consequences both to conquerors and conquered, but the inestimable possession of the Nile.

INDEX

THE END